BEYOND EARTH, THE SOVIET DRIVE INTO SPACE

DECODING THEIR
SATELLITE AND LAUNCH
EFFORTS, 1957-1975:
A VERY PERSONAL VIEW

SAUNDERS B. KRAMER

Proceeds from this title benefit SPACE 3.0, a 501c3 charitable foundation, an organization with the mission to provide grants to projects focused on preserving space history, empowering entrepreneurs and visionaries, and helping to craft a vision for a space future.

© 2025 The SPACE 3.0 Foundation, Inc.

ISBN: 978-1-887022-89-7 (Print)

ISBN: 978-1-887022-90-3 (ebook)

All rights reserved. No part of this publication may be reproduced, distributed, or transmitted in any form or by any means, including photocopying, recording, or other electronic or mechanical methods, without the prior written permission of the publisher, except in the case of brief quotations embodied in critical reviews and certain other noncommercial uses permitted by copyright law. For permission requests, contact the publisher.

Cover Photo. Expertly crafted custom model of the Sputnik 3, an early Soviet satellite that served as a fully instrumented scientific laboratory designed to study Earth's upper atmosphere and space environment. Manufactured at the Moscow Experimental Factory of Visual Aids, the handmade model is constructed of aluminum, steel, and plastic elements. Credit: RR Auction, Space Exploration Auction, Lot 7503, 17 April 2025

The SPACE 3.0 Foundation, Inc.
PO Box 5752
Bethesda, MD 20824-5752
United States
www.SpaceCommerce.org

Table of Contents

About Saunders B. Kramer Sr. vii

Preface 1

Introduction 5

Beginnings 13
    The Early Sputniks   13
    To the Moon   23

Man in Space 33
    Vostok 1   41
    Vostok 2   45

When Are They Launched? 65
    Calculating Orbits   67

To the Planets 77
    Venera 4   95
    Mars   116

Earth Orbit Experiments 127
    Biological Experiments   148

To The Moon—Soft Landing or Bust 155

Soviet Reconnaissance Satellites: 173
    The Fractional Orbit Bombardment Satellite   182

The Orbit Changing Satellites: Prelude to the Orbital
  Destroyer Spacecraft   187
The Maneuverable Satellites: A Second Look   190
The Soviet Anti-Cosmos Defense Force   193

Soyuz 201

Satellites for the People 263
Weather Satellites   273

Zond Spacecraft to the Moon 285

Return to the Moon—A Big Step 301

The Puzzles 329

The Future 337
Interplanetary (Unmanned) Activity   337
Manned Activity—Earth Orbit   358
The Soviet Space Shuttle   361
Earth Orbit—Unmanned   363
To the Planets—Manned   365

Appendix A:
A Calculation for Eccentricity 369

Appendix B:
Letters on *Vostok 3* and *4* 371

Appendix C:
The Injection Velocity for *Verena 5* 375

Appendix D:
The Burn-Time for the *Cosmos 359* Aborted Flight (nee-*Verena 8*) 378

Appendix E:
NASA Prediction Bulletin 381

Acronyms 383

About the Publisher 388

Tables
Table I: Total Launchings by Year 8
Table II: Approximate Time Interval, Δt, for the Arc Length from the Zeroth Node to the Location of the Launch Base for Each Base and for Each Inclination 76
Table III: The Attempts of the Soviet Union to Explore the Planet Mars 125
Table IV: Launchings of Photoreconnaissance Satellites 178
Table V: FOB Launchings 184
Table VI: Altitude Changing Satellites 189
Table VI-A: Altitude Changing Satellites 192
Table VII: Orbital Values for the *Cosmos 373*, *374*, and *375* Spacecraft 198
Table VIII: Launches of the Spacecraft Soyuz 258
Table IX: For 1967-1975, Flight (Spacecraft) Hours and Man-Hours for the Flights of the Soyuz Spacecraft 260
Table X: Tests of Soyuz Spacecraft under the Cosmos Appellation 261
Table XI: The Space Stations Called Saliut—in Orbit from 1971 through 1975 262
Table XII: *Luna 22* Lunar Orbit Parameters 326

Figures

Figure 1: Trace of Flight Path of *Vostok-2* Spaceship..........49

Figure 2. The Flight Paths of *Vostok 3* and *4*..........52

Figure 3. The Spherical Triangle That Describes the Δt from the Zeroth Node to Liftoff..........72

Figure 4. The Two Major Launching Techniques..........78

Figure 5: Diagram of the location of the domain photographed by *Zond-3*..........91

Figure 6. The Capability of Selected Soviet Boosters to Supply Velocity to the Payloads Indicated..........344

Figure 7. Space Station system patented by the author...366

Every resemblance between actual events and characters displayed in this book is intentional. None of the events of times past are fictitious. Many of those discussed as events expected to occur in the future probably will come to pass.

S.B.K.

Dedication

To Yuri Alekseyevich Gagarin and Vladimir Mikhaylovich Komarov, who did so much to point man toward a new future, and to my sons, Saunders Bernard Jr. and Peter Adrian, who will see that future.

About Saunders B. Kramer Sr.

Saunders Kramer's passion was space exploration. "Sandy," as he was known by family, friends, and colleagues, wanted to be the first man on Mars and dreamed of the possibilities of life in space. Born October 30, 1920, he died on May 30, 2005 as he recovered from heart surgery.

A native New Yorker, he called Brooklyn home until moving to California in 1955 where he established his position in the space community very early in the Space Age at Lockheed Missiles & Space Division. At Lockheed, he served as the director/manager of the first detailed study of a crewed space station (1958) followed by space stations for the US Air Force. For a new space station design in 1960, he received the first patent ever issued for one in the US—and likely in the Western World. He is also responsible for the concept and design of the first crewed space tug as well as early space shuttle concepts (1958-1960).

At Lockheed, he was highly involved in conceptual designs of lunar and planetary bases and numerous other NASA-sponsored

Saunders Kramer, 2003
Credit: Elissa Kramer

studies aimed at lunar and planetary exploration. He also taught a graduate-level course at LMSC in advanced systems engineering and design for three years that received university accreditation.

While at Lockheed and until the time of his death, he collected, analyzed, and interpreted data on all satellite/spacecraft launchings, evolving into an extensive database that focused on Soviet/Commonwealth of Independent States/Russian activities. Sandy was widely recognized as an international expert and consultant on this subject.

Sandy was also associated with very early studies on the Polaris submarine missile; examining NASA-oriented missions involving the use of Lockheed's upper stage Agena booster with the design of ballistic missiles; and with early studies in the anti-ballistic missile arena. As a result of his expertise in Soviet space activities, he was a consultant to the CIA and the US Air Force at Lockheed. He also gave almost 200 lectures on many aspects of space activity, both domestic and foreign.

Following his interval at Lockheed, Saunders joined the US Department of Energy in 1971 and conducted automotive engine R&D directed toward reducing exhaust pollution and improving engine efficiency. He retired at the end of 1992.

Saunders served as a contributing editor for *Air & Space Magazine*, was a founding member and fellow of the American Astronautical Society, a fellow of the British Interplanetary Society, a charter member of the Planetary Society, and a life member of the National Space Society.

Publications: He is the author of some two-dozen papers and co-author of another dozen. He also co-authored an early children's book on the Discoverer satellite. In 2003, he authored and, published, *The Hundred Billion New-Ruble Trip: A Russian Landing on Mars*.

Education: Polytechnic Institute of Brooklyn. BS, Physics (1951); MS, Applied Mathematics (1952).

World War II Service: US Army, private first class; honorable discharge in April 1946. Medals received: Campaign Medal, European African Middle East Campaign Medal, Good Conduct Medal, and World War II Victory Medal.

* * *

The following memoir was written by Saunders "Sandy" Kramer in the 1970s and 1980s, with data through 1975. It was never published until now. A profound thank you to Elissa Kramer, the wife of Sandy, for donating this manuscript to the SPACE 3.0 Foundation.

PREFACE

It was 5 AM Pacific Daylight Time, give or take a few minutes, on the morning of October 5, 1957, and the night was quiet as nights should be. The California air was cool but not cold, a sweatshirt provided all the warmth that I needed. No one else in the neighborhood would be up and about at this hour, especially on a Saturday morning. And then I heard the sounds, softly but distinctly, nevertheless, "Do you see it? Do you see it?" "No, do you?" It wasn't one voice but many—within seconds I realized that the same words were issuing from every backyard patio on our long block. I cocked my ear and now could hear the whispers coming from many houses away. "Do you see it? Do you?" Though my ear was tuned to the whispered conversations, my eyes were carefully aimed at a slot formed by an antenna on one house and the chimney of the adjacent house, both outlining a slice of the night sky to the northwest.

Without any advance warning a very bright and swiftly moving star popped up above the horizon. The backyard voices suddenly grew louder, a chorus, "THERE IT IS." And so it was, the rocket case, the orbital stage from the *Sputnik I* satellite booster, silently moved across the sky. I put my binoculars to my eyes, looked behind the booster stage and sure enough a faint but equally swift star trailed the booster some seconds behind it. When I attempted to see the *Sputnik I* itself without using the glasses, I lost it and failed to recover it with the binocs. Muttering to myself, I quickly turned and followed the bright stage; it had passed the zenith and headed for the southeast. When it was still about 50° above the southeast horizon a rather large meteor slammed across the sky under a full and bright Moon. That was nature's greeting to the first artificial satellite. The red-orange ash of the meteor vanished quickly and the booster-satellite followed soon after beyond the horizon. I kept staring at the point where it had disappeared...the world, I realized, would never be the same.

I recalled the events of the prior day commencing a few minutes after 4 PM: I had just left my desk to chat with a colleague on our current studies when I was intercepted by still another colleague, John Cladis, who stopped me to say, "Sandy, the Russians have put a satellite in orbit." I replied, "Aw, John, it's too late in the day for such jokes." He retorted, "No, Sandy, I'm not kidding, Jennie (his wife) just called me, it's true, 'they' put a satellite in orbit." There was no amused look on his face and it dawned on me that this was no jest; I looked at my watch—it was 4:12 PM—I wheeled about, my original intended discussion quickly forgotten, and headed for my boss's office, stuck my head in the doorway and said, "Charlie, the Russians have a satellite in orbit." My tone and the look on my face must have been immediately convincing for his jaw went slack and with great dismay he said, "They did?" The news spread like wildfire. Work ended for that Friday, October 4, 1957.

On the 5 PM news on CBS we heard a recording obtained by RCA at their Riverhead, Long Island, installation. The historic beeps came through loud and clear followed by an announcement that the satellite would be visible the next morning at 5 AM in the San Francisco Bay Area.

From that October 4th until today, I have watched, studied, written, and made predictions about the Soviet space program; that was more than 31 years ago. As for the predictions, I found that once in a while I was correct, but more often than not surprised with each new venture in the Soviet drive to explore the universe.

S.B.K.

Introduction

The Soviet Union has realized, as the US government has not, that the space program is useful far beyond the engineering and scientific advances that accrue because of it. Some think that the Cold War vanished to a large extent during the reign of Nikita Khrushchev but it didn't; it simply was ritualized by virtue of the competition in the space program and during his time in office, the Soviet Union lorded it over the United States for all it was worth. Here was an area where the Soviets displayed a primacy such that even second place—held by the US—was far behind. This was nurtured with care for all the rest of the world to see.

A committee was informally assembled, when [John F.][1] Kennedy was President, largely of economists to examine the effect of the Soviet leadership in the space arena. The major item that I remember in this committee's conclusions was that the US lost some $10 billion worth of business to the Soviet Union in three years (roughly December 1957 to December 1960) due to the Soviet preeminence in space activity.

---

1   Editorial notes and explanations are given in brackets.

The Soviet drive into space, covering all of its multifaceted branches, has shown, first of all, perseverance. One, two, or a dozen failures in some particular area was not about to deter them; no cancellations, only postponements. More about that topic later. After the end of World War II, the Soviet Union was in a state of utter devastation. The cruelties of war had visited upon her so many casualties and so much economic ruin that those who gave it any thought at all were convinced that it would take a generation for anything like normalcy to return along with progress in the areas of the various technologies.

Of the American public who thought about it, the vast majority could only remember a [historical] Russia ruled by the czars and whose people were either primitive farmers or poor workers in the cities, now exploited by a communist cadre that overthrew the greedy, careless royal house. To have explained that the Russians had a long history in rocketry, albeit scattered and thinly supported, even as [physicist Robert H.] Goddard's efforts were in the US, would bring nothing but expressions of disbelief. Most of the American public remain unaware that nonlinear mechanics, a really sophisticated branch of mathematics and physics, which is utterly basic to a scientist's understanding of orbital mechanics and rocket flight, is presently comprised of work laid down by Russians. The names of [mathematician Nicolas] Minorsky, [physicist Aleksandr] Andronov, [C. E.] Chaiken, [mathematician Nikolai] Kryloff, [mathematician and physicist Nikolay] Bogolyubov, and [mathematician Lev] Pontryagin, to name a few, are part of the elite who established the foundations of modern, nonlinear mechanics from which almost all present texts are derived.

By the time a generation had unfolded after the war, progress in the Soviet Union could be measured by their having initiated the Space Age and by having advanced to the point of launching both men and a woman into Earth orbit for days at a time.

[Explorer Christopher] Columbus, though well remembered after 483 years, is in need, more and more, of advocates to protect his title as discoverer of America as other claims are made. In the present world, and for all time to come, none will ever dispute the fact that the first man of Earth to look down on the planet from an orbiting spaceship was Hero of the Soviet Union, Pilot-Cosmonaut Yuri Alekseyevich Gagarin. Gagarin's flight was an epochal engineering and scientific achievement, a real triumph for the Soviet Union. "Probably more in the Western world than in the Soviet Union itself, was it realized that Cosmonaut Gagarin had begun a new chapter in history, one in which man had dared to and succeeded in crossing the threshold of the universe!"[2]

Man's endless dissatisfaction with staying put; his digging and his mountain climbing, his searches under the sea, and his burrowing into the Antarctic ice on this Earth, though far from complete, has a foreseeable end. No matter that it may be some centuries hence when it is accomplished; an end can be foreseen and that's the point. But the frontier that has no end, that will exist for millennia beyond sight or thought, that will demand and receive a sophistication unconceived of at present, not even dreamt of in the most fertile imaginations, has been provided by the initiation of the Space Age. That a Soviet satellite and a Soviet citizen were the foundation stones to everywhere in the universe is a part of history, less immutable that the present laws of physics and the mathematics of Kepler that explain how and why satellites and space probes perform as they do.

Despite the fact that the United States accepted the challenge of space led by the perceptive vision of President John F. Kennedy, the Administrations that followed his, failed utterly to grasp the implications of the infinite frontier and limitless possibilities beyond Earth. The Soviet Union though coming in second in landing men

---

2   *New York Times*, R. H. Anderson, March 29, 1968.

on the Moon—as she will one day before long—has recognized the potential in space exploration; the boundless ea with an infinity of shores. With her inherent and powerful desire to be able to say, "For the first time in the world the Soviet Union has," the Soviets have pressed on undaunted by our marvelous Moon landings. As shown in Table I, the Soviets have surpassed the United States in the total number of launchings ever since 1967.

TABLE I: TOTAL LAUNCHINGS BY YEAR[3]

	United States	Soviet Union	France	Italy	Japan	China	Australia	Great Britain
1957	0	2						
1958	5	1						
1959	11	3						
1960	16	3						
1961	29	6						
1962	52	20						
1963	38	17						
1964	57	30						
1965	63	48	1					
1966	73	44	1					
1967	57	66	2	1			1	
1968	45	74						
1969	40	70						
1970	28	81	2	1	1	1		
1971	30	83	1	2	2	1		1
1972	30	74		1	1			

---

3 A launch is deemed as credited to a country when the launching has occurred from a country's territory through the use of a launching team belonging to that country. The commonwealth relationship makes the British launch an exception, the only one.

	United States	Soviet Union	France	Italy	Japan	China	Australia	Great Britain
1973	23	86						
1974	22	81		2	1			
1975	27	89	3	1	2	3		

From 1967 onward, the ratio of Soviet launchings to those of the US are greater than one but never as great as two until 1970; for 1970, the ratio is very close to three as can be seen from Table I. No matter that many of the launchings of both of the major space powers are military in nature, largely reconnaissance satellites of one sort or another; as the years go by there will be an increasing number of launches for science and exploration and the inevitable prestige commensurate with each scientific advance. Military satellites will reach a point of saturation and level out at some point to keep the armed forces of both countries content. As capability and allotted resources increase, so will the arena of space activities enlarge.

Some say that space activity is a good substitute for war and that it would be well for the Soviet Union and the United States to compete fully, thereby concentrating less on the enormous and wasteful funds for arms. The payoff would be a gain for all nations. So be it. The pressure by the Soviet Union in this area is inexorable as is her desire to lead and to be fully recognized as the leader. There is no evidence to indicate any slowing down in the foreseeable future. Given the unlimited opportunities available, one should expect the frontier of knowledge to be expanded by the Soviet Union, year by year, as she moves farther and farther into space—into the outer reaches of the solar system and beyond.

Somewhere history will reveal, one hopes, the persons who recognized the possibilities in spaceflight and convinced the Soviet political leaders of the gains that lay therein. An even more astute political leader—was it Krushchev?—realized the potential from his point of view and supported the plans of science and engineering to enhance the fortunes of the Soviet Union in the world. Those key decision-makers have forever redirected man's future. Before the 1900s are part of history, man will have had a look at every planet in the solar system whose existence he is aware of (there are thoughts about a planet beyond Pluto but, to date, it remains undiscovered).

Though in comparison with other ages, the Space Age is but a newborn babe, it has brought forth as varied a brood as man's excogitation could conceive; of course, the period of conception goes on forever. The Soviet Union's scientists and engineers were concerned with the possibilities of human spaceflight very shortly after World War II. Geophysical rockets that carried payloads in almost vertical paths to hundreds of kilometers lifted thousands of pounds of instruction and instrumental animals in biological research efforts commencing in 1951 and extending to the present day.[4] These verticals, along with research efforts in countless other places, led to the age of space. Because of their initial biological studies with the verticals, the Soviet scientists were prepared very early, indeed, in their space activity to launch biological specimens into space, as witness the female dog, Laika, in *Sputnik II*, launched on November 3, 1957.

These biological flights are inextricably tied to manned launchings. Close to these are scientific flights to determine the nature of the space environment, both near and far, and to supply new knowledge to the esoteric world of physics. It is but a short conceptual step

---

4 One of these vertical launches, a multicountry effort, was sent up on October 3, 1970, at 0420 GMT to an altitude of 500 km and was instrumented for a series of studies of the Sun.

then to meteorological satellites to look at our own environment on Earth. And if weather satellites, why not satellites to look at Earth resources; all of them? Since we are a resource unto ourselves, it follows that we use space to improve that use, ergo communications satellites. If the civilian aspects of our lives can be aided by metsats [meteorological satellites] and comsats [communications satellite], well, so can the military, and thereby were born military comsats and navigation satellites that will in turn also find civilian application in the multitude of aircraft flights.

That innocuous euphemism, "national means of verification," meant still another military addition to the satellites' spectrum—reconnaissance, both photographic and electronic—which comprise a large part of the Soviet space activities and about which very little is known. When discussing military satellites, one cannot neglect the FOB (fractional orbital bomber), which despite chatter to the contrary is actually a satellite—that is, it does attain orbit and is not just a long-range intercontinental ballistic missile (ICBM), which, by the way, just snoops around at orbital altitudes for a short while but is by no means a satellite. During the time when, as Secretaries of Defense will, he defended the posture of the US as not needing, but able to produce them, Robert McNamara, in a brief phrase, revealed much about the nature of the Soviet FOB.

The most interesting aspects of space activity in the Soviet Union, paralleling that of its human spaceflights are its efforts in interplanetary flight, including flight to the Moon. The Soviet Union has spared no effort in this broad endeavor. Starting as early as 1959—there were rumors that some unsuccessful attempts were made even earlier—spacecraft-based studies of the Earth's present natural satellite were undertaken with the Luna series of probes. This was followed by the Zond lunar series commencing in 1968. The Luna series can be expected to continue for some time. In October 1960, the Soviet Union tried to launch two spacecraft to

Mars. These were the first in a long series of unsuccessful attempts to reach the planets. Venus flights commenced in February 1961—the scientific community is pleased with the successes to date in that effort. Few attempts have been made to reach any other planet, and other planets are considerably more difficult to reach.[5]

Ever since the Space Age began, I have kept records of the action by all of the countries involved but my main and never-ending interest lies in the launchings by the Soviet Union, not as launchings per se, but in the various experiments and missions conducted using a spectrum of Earth satellites, manned spacecraft, lunar/probes, orbiters, and landers and interplanetary spacecraft. I am not a casual observer of space activity for I've collected data, analyzed, and studied man's efforts in the extraterrestrial arena from day one of the Space Age, October 4, 1957, until the present, and will continue to do so for the foreseeable future. Outside of the intelligence communities of whatever countries indulge in such activity, there are few who have followed the initial tentative steps and the increasingly imaginative and bold explorations by the space powers as I have.

To the phrase that nothing is certain but death and taxes one can add, "The Soviets will reach out for all of the solar system and beyond;" and because of their perseverance, quite beyond belief, they will succeed.

---

5   Soviet efforts to the end of 1975 have been confined to Mars and Venus. The United States has initiated exploration of Jupiter and Mercury.

1

BEGINNINGS

THE EARLY SPUTNIKS

Sputnik I was launched from the Tyuratam launch base near the Aral Sea at about 2100 GMT (Greenwich Mean Time) on October 4, 1957, into an orbit with perigee at 141.7 statute miles, apogee at 588.4 statute miles, inclination to the equator of 65.1°, and with an orbital period of 96.17 minutes. The uproar that followed the event is all part of history and one would find it perhaps difficult to believe today in view of the blasé attitude the public has assumed in recent years. A good portion of the ho-hum attitude is due both to the success in the launching, by all the nations involved, of over 1500 satellites and space probes since (to the end of 1975). And to the lack of real understanding—very widespread—in the Soviet Union, as well as the rest of the world, as to the complexity and sophistication of the systems involved and the capabilities and dedication of the people in the space community.

The public everywhere is turned on by something new but, being a species whose appetites are quickly jaded, large percentages of success weaken public support rapidly. Almost as rapidly, frequent failures lead to dismay and a tendency to wander off in directions leading to adventures with a reasonable taste of success early in the game. As in every other human endeavor, reasonable success (whatever that means) together with sharp competition constitutes a satisfactory formula assuring continued interest.

When the weight of *Sputnik I* was announced as 184.3 pounds, many were quick to suggest that the information was incorrect and that a decimal point, no doubt, had been misplaced somewhere in the original TASS release; if *Vanguard* was to be 21 pounds, then "obviously" the Soviet satellite ought to be 18.3 lb, of course!

The TASS release gave the weight, the diameter as 23 inches, the shape as spherical, observed the presence of four antennas, two of which were 2.4 meters long and two at 2.9 meters (7.8 and 9.5 feet, respectively). They were displaced from each other by 90°. The radio frequencies chosen, 20.005 and 40.002 MHz,[6] were selected so that they would be ducted more than 180° around the Earth by the ionosphere. At these wavelengths—about 15 and 7.5 meters, respectively—radio waves are reflected in the ionosphere as though trapped in a tube and are thus carried long distances making it much more convenient to tune in on the satellite as every "ham" worth his salt did while *Sputnik I* continued to broadcast.

Broadcasting ceased on October 27. If the one-watt power output, supplied by the batteries, for the transmitters (each) sounds small to the uninitiated, it is worth noting that for some time early US satellites had transmitter output power measured in small fractions of a watt. The Soviets have always had more than adequate power for all of their spacecraft. The satellite had instrumentation that permitted detection of temperature and temperature changes, and some measure of the Earth's magnetic field strength and little else.

---

6   Megacycles are old hat; Hertz is the discoverer of radio waves.

Nitrogen gas, under a pressure of one atmosphere, was circulated in the sealed sphere as a heat regulation device.

Until this launching, investigations of the properties of the upper atmosphere were carried out using sounding rockets (vertical launches) that stayed at the altitudes of interest for just a few minutes. Satellites offered the opportunity to measure such properties for days. As long as the transmitters operated—23 days—temperature could be studied. Tracking the satellite and its booster stage with or without radio transmission permitted experimental determination of the density of the upper atmosphere by observing the decay (i.e., the change in altitude and orbital period due to the energy loss of the satellites because of collisions with the molecules of "air" remaining at those heights).

Actually, the American press did not record much information concerning instrumentation and the results thereof but speculation abounded. The Soviet passion for secrecy and omission of details, particularly on the booster used, fed the speculation, and this author didn't avoid all the brambles either. My early analysis developed a three-stage booster (after all wasn't that the configuration for Explorer and Vanguard?); the facts did not come until years later with the revelation of the Vostok booster details. The actual booster configuration for *Sputnik I* presented a strange sight to the eye since it consisted of the Vostok booster without the upper stage so that it had a truncated appearance.

The booster was a direct outgrowth of the Soviet ICBM with the weight of the warhead largely "exchanged" for the additional velocity necessary to reach orbit. This configuration was also used to place both *Sputnik II* and *III* into orbit. Despite the success of this event the Russians, it appears in retrospect, were not as prepared as they should have been to make maximum political and propaganda use of the launching. It must have been an interesting dilemma since they were anxious to make as much out of the launching as possible; particularly to point out that we, the US, were in second place while the fog of secrecy that controls their press prevented a display of details.

The booster as a military vehicle was kept under wraps for years until, as an ICBM, it became displaced by smaller ballistic missiles more suited to the military needs. To this day, it is a great workhorse of the Soviet space program.

Sputnik I performed 1380 orbits and traversed about 60 million kilometers (37.2 million miles); it decayed from orbit on January 4, 1958, having lasted 92 days. The booster, the object that was really seen by the public, traversed some 39 million kilometers (24.2 million miles) and came down on December 1, 1957—some 57 days after launching occurred. Its debris probably fell on Alaska and along the West Coast of North America.[7]

If the American public along with the rest of the world was startled with the launching of *Sputnik I*, the launching less than a month later of *Sputnik II* was a thunderclap of Olympian proportions. Here was no package of instruments but rather a sophisticated representative of Earthian life in orbit around its mother planet. Laika, a female dog was instrumented to measure the animal's pulse, respiration rate, blood pressure, nervous system activity, and muscle motions. The cabin was instrumented to determine temperature and pressure and had an air conditioning system to aid in satisfying the life support requirements for the dog. The launching of Laika into orbit raised a howl of protest from dog lovers 'round the Earth. Letters of protest appeared in major newspapers everywhere. These same protesters did relatively little to curtail research using animals in their respective countries but this display on an international scale of

---

[7] "Soviet Artificial Satellites Space Probes and Spacecraft," Biull, Stantsii, Optich, Nabliud, Iskusst, *Sputnik Zenli*, no. 45, 1965, pg. 42.

sacrificial biological research outraged their sensibilities. The Soviet Union quietly ignored all the background noise and proceeded with its pacesetting experiment.

Sputnik II was in an orbit of 1,038.3 by 139.8 statute miles at an inclination of 65.3° and an orbital period of 103.75 minutes. It was launched at 0232 GMT on Sunday, November 3, 1957, again from the Tyuratam launch base. The experiment itself was short-lived, lasting just a week during which information on Laika's behavior in orbit was carefully recorded. Besides the biological experiment there were instruments to investigate ultraviolet and X-ray emissions from the Sun and cosmic rays in general.

By this time some dozens of researchers in many countries were evolving curves to permit determination of orbital lifetime of a satellite, given its perigee altitude, its mass in orbit (its weight on Earth), its shape, and its orientation in orbit as it "flew" around the Earth. The Soviet Union had given out enough information on *Sputnik II* to permit such curves to be used with reasonable expectation of getting results to accuracies of no worse than 50-100 percent (i.e., the lifetime could be determined to within a factor of two or less). Breakwell and Koehler[8] set up such a set of curves and, since they were friends of mine and also colleagues, I took a set and commenced playing the game of applying this to the lifetime of *Sputnik II*.

My calculations said that the satellite would decay after 2,370 orbits about the Earth. In essence then, using the uncertainty factor that we all thought was valid, the satellite should have stayed up at least 1,185 and not more than 4,740 orbits. The Soviet Union had announced that the totality of experiments had not been separated from the orbital stage of the booster; consequently, when decay

---

8   "Elliptical Orbit Lifetimes," J. V. Breakwell and L. F. Koehler, *Advances in the Astronautical Sciences, vol. 3, 1958, Plenum Press, New York.*

occurred there would be a good chance that it would be seen by some peoples on Earth. Such a large stage—it was estimated to be 70 to 80 feet long and 8 to 9 feet in diameter—would form a brightly glowing swath many miles in length as the orbital energy was dissipated in friction with the atmosphere.

On April 15, 1958, at 8:55 PM Eastern Standard Time the satellite reentered over the Lesser Antilles (Martinique, Antigua, etc.) and was seen in the Caribbean area and in the Guianas and in Venezuela. One of many descriptions available estimated the glowing reentry to cover a swath at least 120 miles long. The reentry took place on orbit number 2,370. Now, the upper atmosphere varies over a greater spectrum than does the lower atmosphere in which we live and breathe. Activity on the Sun is largely responsible for this phenomena and the variation can be as great to cause the drag on the satellite to change by as much as a factor of four.

In 1957-1958 we were a long way from getting close to the actual forecasting of these upper atmosphere changes. I was considerably more surprised than the reader of this book might be in getting so accurate a result. In spite of the odds that appeared to exist on making such predictions for future satellites, I was able, in the future, to repeat that kind of performance several times to the great entertainment of my colleagues. Since I find it hard to believe that I was able to make calculations that well, I have long since decided that my lack of knowledge about the satellite shapes, together with the uncertainties in the upper atmosphere density at any given time, form a smoothing function that tends to cancel out any large errors in the calculation. Well, I can't find a better explanation.

The orbital inclination of *Sputnik II* permitted a study of the Van Allen radiation belt especially near the Earth's magnetic poles despite the fact that the radiation belts weren't fully realized (or named) at that time. Frequencies used in transmitting information were again in the 40 and 20 MHz band, selected for the fact that they were well transmitted by the ionosphere waveguide effect used so effectively in the case of *Sputnik I*. A new and related effect appeared as well. There were "signal predecessors," which appeared some 30 seconds before the beginning of main reception from the satellite, it was a convenient phenomenon.

Of course, the main output from this experiment was that man would be able to survive in space in a condition of weightlessness for a week—if flights that long should be forthcoming. Few recognized that manned flight was not very far off, we would wait a miniscule interval, even in the history of man let alone cosmological considerations.

On May 15, 1958, the third Soviet satellite appeared in the skies and again startled both the public and the scientific world by virtue of its size. The *Sputnik III* satellite weighed 2,925 lb and carried 2,134 lb of instrumentation, which represents a remarkable efficiency since 73 percent of the orbited weight is devoted to obtaining scientific information while the remainder of the satellite was in effect a big "rack" upon which to hang all of the instruments in selected orientations and locations.

Second-guessing by both the lay public and the scientific and engineering community led to widespread opinion that *Sputnik III* was heavy because of the use of instrumentation generally found in Earthbound laboratories (and therefore built without a concern for weight). Nothing could be further from the truth. Despite the fact that some instruments were of the vacuum tube type, the use of solid-state instrumentation was the order of the day as one could determine by reading the papers on the experiments and by reading, no less, the dryly factual reporting of TASS, the Soviet news agency.

Physically the satellite was large, 140 inches long and with a diameter of its base at 68 inches; it looked like a cone with numerous protrusions, which were antennas and various instruments. The orbit was 1168 by 140 statute miles having a period of 106 minutes and again at the inclination of 65°.

The orbit, the weight, the size, and the impressive list of instruments indicated at once that this was to be a long-lived, detailed, scientific endeavor and that planning for this launching had begun a long time ago, certainly before the launch of *Sputnik I*. It was not to be until 6 years 3½ months later that the US on September 5, 1964, would launch a similar orbiting geophysical observatory (OGO) for that is exactly what *Sputnik III* was. The workmanship was superb as evidenced in a full-scale model exhibited at the Brussels World's Fair in 1958. The experiments included investigations of micrometeorites, composition and change in intensity of cosmic rays, recording of heavy element nuclei in cosmic rays; measurements of solar radiation and the electrostatic field of Earth, electrical charge on the satellite, the content of the ionosphere, the magnetic field intensity, the temperature of the upper atmosphere and of the satellite itself. A beacon radio transmitter aboard *Sputnik III* operated for the full 6,500 hours of the satellite lifetime.

Power for the satellite was supplied for 4,000 hours by batteries and for the remainder of the time from solar cells. The satellite remained in orbit for 692 days—10,037 orbits (I had calculated 930 days) and decayed on April 6, 1960.

The first three satellites yielded much information—the Earth's oblateness was determined to a greater degree than had been obtained in all the prior years of measurement, the density of the upper atmosphere was discovered to be almost 10 times greater than had previously been supposed, and its variation with solar activity was observed and measured. (Increased solar activity caused the atmosphere to expand, thus increasing the density at higher altitudes and in turn lessening, then, the lifetime of orbiting spacecraft—the lifetime of the third satellite was calculated by the Soviets based on results collected during the lifetimes of *Sputniks I* and *II* when the Sun was active. And, they state, that *Sputnik III* far exceeded lifetime expectations since by the time it was launched any solar activity had calmed considerably, thus lowering the density at its orbit altitudes)—the predomination of oxygen ions in the mid-latitude range (between 28° and 65°) at altitudes above 140 miles and continuing past 300 miles, the nature of the variations in and the strength of the Earth's magnetic field at orbital altitudes was determined and measured for more than a year, the radiation belts were studied and their contents identified and measured, cosmic rays with heavy nuclei were measured and it was found that the proportions of heavy elements in the rays were essentially the same as found elsewhere in the universe, finally (and happily) it was shown that the meteorite danger had been grossly exaggerated. Now, as history has recorded, there were no further Soviet launches (at least that were successful) until January 1959. It remains difficult for me to believe that riding on the crest of a wave of success that the Soviets sat idly by for seven months. If indeed, failures did occur, they have been well hidden from the public.

The gap in the Soviet space program also led to all sorts of speculation and some of this provided considerable amusement and not a few real believers, even in the "space community." The following article appeared in the Swiss newspaper *Basler Nachrichten* sometime during the month of December 1958.

> The Soviets have set the month of May 1959, as the date for launching the first "sputnik skyscraper." An oval, cucumber-shaped space station will revolve in its orbit at 4,000 kilometers altitude (2,485 statute miles), will measure 330 meters in length, and have a maximum

diameter of 85 meters (1,080 feet long, 280 feet in diameter). One hundred and fifty transport sputniks are now under construction. They will place in the chosen orbit both the materials necessary for the construction of the supersputnik, which will be constructed on the spot, and 75 "working cabins" necessary for assembling the supersputnik. The research on these cabins is already completed. Professor Blagonravov is credited with the statement that the problem of returning the space workers to Earth has been solved. No worker will remain on the job more than 5 days. The regular crew for the completed "skyscraper" will consist of five men and will be relieved every eight weeks by means of round-trip rockets. Scientists from all countries will be invited in groups of five to visit the "sputnik skyscraper," named Robot I.

This article also appeared in English in an American journal known for its sophisticated technical articles; I'll spare them the embarrassment of identification. It should be evident from the article that the pastime of smoking pot is hardly a recent innovation, the writer surely was high on something.

TO THE MOON

In December 1958 I wrote an article for the magazine *Aviation Week* (now called *Aviation Week & Space Technology*) discussing the subject of trajectories to the Moon. At the end of the article, I observed that the month of January would be an appropriate time for a launching to the Moon by the Soviets. Unfortunately, my powers of prophesy were not used to the best advantage for the article was misplaced and not published until the January 19, 1959 issue appeared.

Luna I was launched, with the explicit intent of impacting the Moon, on January 2, 1959, at about 1700 GMT. The launching encompassed the use of a direct trajectory; that is, there was no intermediate path in an Earth "parking" orbit. The technique involved aiming at a point where the Moon would be at the same time that the interplanetary probe arrived at that location. Because there were no auxiliary engines aboard the last stage of the booster that contained the payload, it was not possible to make any course corrections en route to ensure impacting the Moon. As a consequence, and because the booster guidance had a small error either in "aiming" or in burnout velocity (or both), the probe missed the Moon by about 3,100 miles, approaching that close to the lunar surface some 34 hours after launching at 0600 GMT on January 4. The error was, angularly speaking, very close to one degree assuming that the impact point intended was the center of the Moon. Burnout of the last stage occurred within some 120 miles of the Earth's surface; within a few hours of tracking the probe the Soviets knew that they had missed their target.

Although the prime mission was not accomplished, *Luna I* obtained new data on the radiation belts, made cosmic ray end meteoroid flux measurements, "searched" for a lunar magnetic field (it found none of consequence) and conducted other related experiments. Information on the experiments was broadcast to Soviet receiving stations (along with anyone else who had the antennas to listen in—e.g., Jodrell Bank at Manchester, England) on a frequency of 183 megahertz (MHz). The probe also broadcasted using frequencies of 19.993, 19.995, and 19.997 MHz, respectively. The probe was tracked by the Soviets for 62 hours out to a distance of 373,000 miles, the longest range at which any spacecraft had been tracked to that date.

Tracking ceased because the batteries aboard were depleted, not because of the long range. The expended last stage and its payload weighed 3,245 lb of which 796 lb was the probe *Luna I*. At 0056:20 GMT on Saturday, January 3, when the stage was 74,400 miles from Earth, an artificial comet was created by ejecting 1 kilogram (2.2 lb) of sodium in its atomic state from the rocket. The reflected light from the sodium cloud had a brilliance of a star of the seventh magnitude, just too small to be seen by the naked eye but easily viewed through a low power telescope (if, of course, one knew where to look).

Luna I went into a heliocentric orbit on January 8 with the parameters: perihelion, 91.5 million miles; aphelion, 123.25 million miles; a period (i.e., its "year") of 447 days; and was inclined to the ecliptic by less than a degree. So was born man's first artificial planetoid. So did the fascination with space activity increase in the public eye.

The remainder of 1959, until autumn, was a period of apparent quiescence for the Soviet space program and once more one wondered if other launches were tried in that period. However, I, for one, have never turned up any convincing evidence to indicate that any launching attempts were made in that interval. During those

Chapter 1: Beginnings

eight months before the launch of *Luna II*, Soviet engineers and scientists carefully analyzed not only the data from *Luna I* but also critically examined the guidance-error problem that led to the lunar miss of that probe.

Once again, on the morning of September 12, 1959, at 0700 GMT, the launch pad at Tyuratam came to life and *Luna II* was launched using a direct trajectory to the Moon. The probe, even as *Luna I* was termed, was called *Mechta (Dream)*, but this was a dream come true. *Luna II* plus its carrier—the last stage of the booster—weighed 3,324 lb, an increase of 70 lb over its predecessor; 63.8 lb of this was in the 859.8 lb payload and 15 lb was accounted for in additional equipment on the stage. Indeed, the stage itself, this time had a transmitter and it broadcast on frequencies of 19.997, 20.993, and 39.986 MHz and the third transmitter, from the probe, broadcast information on 183.6 MHz. The same types of instrumentation were aboard as in *Luna I* and again the probe was a sphere, sealed and pressurized with nitrogen to about one atmosphere.

Luna II struck the Moon at latitude 30° north, longitude 0° and about above the northern slopes of the Lunar Apennines according to observations made in Hungary (Budapest and Bayer) and in Sweden (Uppsala). Impact occurred at 2102:24 GMT on September 13, 38 hours 2 minutes after launching. A series of photographs made at Uppsala University Observatory depict a cloud of upraised dust that lasted for 1.9 minutes. The Soviets said that this was in good agreement with their radio fixes of the impact. Cloudiness over the Soviet Union prevented direct observation by optical instruments. On September 12, a second sodium cloud experiment was conducted at 1848 GMT at a range of about 74,000 miles from Earth. The sodium cloud lasted long enough to permit 5-6 minutes of photographs in the Soviet Union.

During the September flight, the cosmic ray detection instrumentation picked up a twelve-fold increase in particles with atomic number 15 (i.e., phosphorus) and heavier elements. Correlation with two solar flares that had occurred a few minutes earlier was made. Since *Luna II* impacted on the Moon it was possible to take measurements of any magnetic field due to the Moon up to a few seconds before that impact; at 50 kilometers altitude a magnetometer aboard provided data to show that any field had to be 400-1,000 times less than that at the Earth's surface.

The probe and last stage impacted the lunar surface at 3.3 kilometers per second (10,877 feet per second). I made some calculations and showed that the kinetic energy released at impact would, if all converted to heat, be enough to vaporize the probe, plus rocket, even if they were made of steel. Since the rocket and probe were largely aluminum alloy the energy was far in excess of that necessary to produce clouds of hot metallic vapor. Of course, all the energy was not so expended; a large part dug out a crater that must be in excess of 100 feet in diameter and 20-30 feet deep that was due to the expended stage, with the probe crater somewhat smaller. The crater formation gave rise to the clouds of "dust" seen by the Hungarian and Swedish observatories. A lesser portion of the energy was expended in deforming the structures of the probe and the rocket stage. Despite all of the energy available in these processes, the impact, crater formation, and structure deformation is a complex affair so that I would not be altogether too surprised to see that some debris remaining in the area of impact could be recognized as having come from the first spacecraft to have struck the Moon. Score two for the Soviet Union!

At the time that these new space events were taking place, the Soviet radio backed up by astronomers and lunar experts expounded at length on what might be found on the lunar surface: opinions included the possibilities of lower forms of vegetation in moist crevasses, oil and gas deposits, and a spongy-porous upper surface layer. The latter has been disproved in at least three locations on the Moon to date. As to vegetation or gas/oil deposits one can only say that, by early 1975, no instances of these have shown up—keeping in mind our very limited exploration of the lunar surface; that no deep subsurface samples have been obtained, and that no crevasses containing hoar frost or other moisture have been found supporting lichens or any other lower forms of plant life. In fact, no crevasses have been found or explored by man or automatic spacecraft to this date (December 1975). It is interesting to contemplate the course of events should such a discovery take place.

Dawn on the morning of October 4, 1959, looked down on Tyuratam and found a crew bustling about with restrained anticipation and brooding concern—this was the second anniversary of the Space Age and a new and historic flight was in the final stages of preparation. The final wisps of vapor from topping off the liquid oxygen tanks vanished as valves closed, umbilicals dropped away, engines ignited, the holding arms lingered until thrust built up, and then, on computer command, rotated away from the booster as the space vehicle lifted off the pad at 0100 GMT (it was 0600 at the launch base, local time). *Luna III* was on its way, initiating a mission whose successful completion would not only startle the world, delight the scientific community including me but also give rise to a stream of invidious articles by some know-nothings in an effort to prove the *Luna III* results false (i.e., that the photographs were fakes, etc.)

Because this spacecraft, the Soviets called it an Automatic Interplanetary Station (AIS), was to return to the vicinity of the Earth, its initial velocity at burnout was less than that of the previous lunar probes by some tens of feet per second to ensure it having an elliptical orbit with apogee beyond the Moon and perigee near the Earth. The orbit—because of the influence of the Sun, the

Earth, and the Moon—changed radically after *Luna III* returned to Earth on its first pass through perigee; however, the initial apogee was at 292,000 statute miles from Earth and occurred at about 1700 GMT on October 10 while perigee was at 24,853 statute miles and occurred at 1650 GMT on October 18. The orbital period on this first pass was about 16 days. The instability of the orbit was such that its perigee shrunk and its apogee lengthened with time until perigee dipped deep into the Earth's atmosphere and the spacecraft reentered and was consumed by the friction with the air in April 1960.

Communication with *Luna III* had ceased in November immediately after recording a substantial meteoroid strike. It should be observed, that at final perigee, *Luna III* was traveling at close to 35,000 feet per second due to the perturbations on it noted above; entry at that velocity assured its complete destruction. At 1416 GMT on October 6, *Luna III* passed at its closest distance to the Moon, 3,880 miles from the surface and then proceeded on its trajectory to the farside, invisible to the Earth. The spacecraft reached a distance of 40,500 miles from the lunar surface at 0330 GMT on October 7 at which time photography was initiated. The angle between the spacecraft and the Sun-Moon line was small, a control program stored in the vehicle was initiated by a timing device (almost certainly backed up by a command from the control center in Moscow or in Yevpatoriya on the Crimean peninsula—about 50 miles north of Sevastopol).

The control system brought the slowly spinning spacecraft to a fixed position with cameras aimed at the lunar farside—the spinning was stopped—and the two long focal length cameras (200 and 500 mm) were placed in operation for 40 minutes. A large series of photographs was made using two different scales on 35 mm film, which then was processed under conditions of weightlessness. The film was developed, fixed, and dried; a single bath solution provided for the developing and fixing in just three minutes. The film is called

"Isochrome" by the Soviets and in addition to its other properties it was heat resistant. Because the processing took place in the environment of weightlessness the viscosity of the developing fixing bath was increased since liquids have a tendency to flow with greater ease under "zero g" conditions.

Prior tests on Earth before the flight were performed using the selected solution and it was exposed to simulated space conditions and temperatures from 86 degrees F to 122 degrees F for 15 days; it worked to great satisfaction under these test conditions. Even when brought up to 158 degrees F the resulting film fogged only slightly. These procedures represented a substantial advance in photography since ordinary film is developed at 60-65 degrees F and is very sensitive to temperature changes. The film used was also exposed in tests to a 12 MeV electron beam from a betatron and this—simulating exposure to cosmic rays in space—caused only insignificant blackening of the film. *Luna III* was designed and constructed to prevent this energy level from reaching its film except, I imagine, under the most violent of solar flares, an infrequent occurrence.

Since the farside of the Moon had never been seen by the eye of man, the Soviets wisely decided to define a base for the coordinates of the areas to be photographed on farside by photographing a goodly number of known areas on the visible side of the Moon. The superposition of the grid formed by working with the "known areas" on the farside led to accurate location of the latter. Moreover, photos of nearside objects would permit a proper assessment of the transmission quality and reliability of the farside surface objects that were photographed (i.e., did the objects really look as they were depicted in the photographs?). One hundred and seven formations were photographed, fifty-one of which were visible from the Earth. The selenographic coordinates were ascertained for all formations and an *Atlas of the Moon* compiled for the reverse side of the Moon. (It is available in the US, in English.)

According to the Soviets:

> An investigation of the photographs shows that the Moon is asymmetric relative to the plane separating it into visible and invisible parts. There are no extended ravines on its reverse side, such as the Ocean of Storms and the large mare. The total area of the detected seas ("Sea of Moscow" and "Sea of Dreams") equals the Sea of Rains on the visible side. There is no continuation of the "belt of seas," although (there is) a group disposition of some details in the structure of the mountain regions and large craters in places. Craters occupy a major portion of the area. The formations on the reverse side do not differ in their nature and structure from those observed on the visible side of the Moon; there are no specific objects. The albedo of many regions is elevated, and the bottom of many craters is dark. Sometimes, a high brightness and, perhaps, luminescence of the center ridges is noted. The porosity (micro-relief) of the surface on the reverse side is the same as, if not more than, the visible side.[9]

In order to assure themselves of receipt of the results of their—and *Luna III*'s—labors, the Soviets had the good judgment to initiate transmission of the images via a slow scan process at the probe's maximum distance—that is, immediately after processing—so that the least possible time would pass in which any unforeseen event might occur that could negate all of the effort expended in this mission. Each frame consisted of 1000 lines (a home TV in 1978 uses only 525 lines) and the frames were probably scanned several

---

9   No reference given in original document for this quoted text.

times during the course of the probe's travels through space to the Earth.

After a frame had been scanned once it was wound up on a reel upon command from the Earth. The process was rerun to whatever extent time permitted; this allowed computer overlay processing to get the best possible picture when all results were in. The televised half-tone pictures were scanned at a much faster rate when *Luna III* reached perigee at the Earth. Recording was made on everything available; on various films, on magnetic tape, on electrochemical paper, and was stored in videotrons (vacuum tubes of a class that stores images in electronic format, which must be "read" out within the storage-time capacity of the tube; otherwise the image eventually "leaks" away).

The photographs were actually taken under difficult conditions, in that lighting from the Sun was essentially overhead for the procedure and as a result the contrast obtained was not all that one could hope for. This led to the need for matching the best photographs from all that were taken and recorded. At that stage of space photography—the beginning of a brand new and complex endeavor—some of the TV lines were seen as separate "objects" in parts of some of the pictures reproduced in the press. This was also due in part to the desire to get these photos before the public, everywhere, as soon as possible.

There are many persons, still today, whose pastime is denigrating the efforts of the Soviet Union without any selectivity, and, equally, without the knowledge to discuss many of the subjects involved. Some of the early flaws in the *Luna III* photos were picked by not a few of these round-heads and were offered to newspapers and magazines alike as "proof" that the photos were all a fake. It's difficult for some people to realize that competitors of the US frequently turn out products earlier or better or both. One such individual collected an impressive array of opinions from persons knowledgeable in photography and its related technology and these covered "experts" from the academic world as well as industry.

My colleagues and I also studied the photos and realized quite early what had occurred. We kept our opinions to ourselves, waiting for the Soviet Union to make available the improved pictures we knew would come. Our discussions, however, within our community were considerable. We noted that none of the authors of articles complaining of fakery had based their conclusions on information obtained from any authority in the US space community. When, finally, the individual referred to above ventured to phone a colleague for an opinion—well after his article had appeared in print—the summation of my colleague's remarks was to tell his acquaintance, and I quote, "You are a schmuck!" So much for that sort of nonsense—but we were to see it again and again, particularly when manned flight commenced. Of course, our own (US) photography has long since confirmed the Soviet findings. As if they needed confirming!

Of all the formations named by the Soviets, the crater Tsiolkovskiy intrigues me the most. Its so-called central peak—really a series of peaks, a small mountain range—reminds me of the Spitzbergen in the Mare Imbrium (the Sea of Rains) on the visible side of the Moon. And the latter is the area that the Soviets appear to have selected for a future lunar base. The surveying initiated there by *Lunokhod I*, landed by *Luna 17*, supports my thoughts about such future activity.

2

MAN IN SPACE

In the months between the initiation of the Space Age and the occasion of the first manned spaceflight, there were all sorts of forebodings both by persons who were in some position of authority—members of the aeromedical community, for instance—by many who on the "basis" of their theological reasoning thought it wrong, and those who opposed it because (a) they didn't understand the principles involved and therefore were certain it couldn't be done and (b) they were part of the general public who were against it, period. (Take any sample population from among the public, pose a question and you will find: 15 percent will approve—or say yes with no hesitancy; 15 percent will say no, no matter what; and the remainder will be persuaded one way or the other depending on time, arguments offered, gains to them, or losses, and so on.)

In the aeromedical community one well remembers [German physicist] Heinz Haber, his lectures supported by animated cartoons, which shoved a man in a spacecraft crushed by acceleration, stabbed constantly by cosmic rays penetrating the spacecraft, boiled by the heat of the Sun baking the craft, and having one hell of a time trying to eat, drink, and perform various tasks. Hand-in-hand with this emotional approach went a letter to the Academy of Sciences of the

USSR signed by some very reputable US medical doctors assuring the Russians that weightlessness would be such a shock to man that he wouldn't survive the event. The second group is best defined by a tale well known but worth repeating: Wernher von Braun after a well-received lecture in the Huntsville (Alabama) area is approached by a sweet-faced, grandmotherly lady who scolds him and says, "Why don't you stop shooting rockets into the heavens and stay home and look at television like God intended you to?" The (a) class of individuals above are of the same hue that told Goddard that his rockets wouldn't fly in a vacuum. The (b) class I'll try to ignore.

There was a really insidious group at work as well during the period before the first flight. These consisted of those "remarkable" humans who have managed to outdo all of the intelligence agencies—in that they have regularly heard via ham radio the cries for help of cosmonauts in foundering spacecraft while the agencies with all their extensive equipment heard nothing. Because of the spectacular nature of spaceflight in the early days, it made marvelous press releases; all of these tales which in retrospect are obvious lies "of purest ray serene," were well covered and the stories became very widespread. There persist to this day many, who being virulently anti-Soviet, really believe that nonsense.

One of the early tales revolves around two brothers in Turin, Italy, who had, on their ham radio, heard "moans from space intermixed with pleas for aid." In view of the fact that they gave a time for this adventure story I made a check of all satellites that were in orbit then—there were very few—and found that sure enough one of the US *Explorer* satellites was well within radio range of Turin at their quoted time. Listening to its telemetry signal converted to audio range from a tape I obtained (its telemetry broadcast on 1.8 and 19.999 MHz) I heard a distinct rising and falling sound that sounded very much like a moan.

A second tale concerns the presumed orbiting of three cosmonauts, one of whom was a woman. In this case, all control was supposedly lost and the intended Earth satellite went off to the planets and the cosmonauts were subsequently lost. Going off to the planets requires some 13,000 feet per second more velocity than orbiting the Earth. Both the space powers in the embryonic Space Age were happy enough to get their satellites into Earth orbit without the burden of carrying more than half again as much propellant for no evident purpose. The Soviet booster was large but not quite that large…or wasteful!

My wife and I had a dinner engagement at the home of some friends on Saturday evening, May 15, 1960. We were discussing the pleasant prospects that seemed implied in the fact that my company was going to apply for a patent for a manned space station design I had developed. I was basking comfortably in the afterglow of publicity that accompanied my presentation of the design to what was up to that time the largest technical audience ever assembled to hear a paper in the space community.[10] The possibilities for the US space program welled up like a Niagara of serendipitous adventure. In that mood we turned to the television set for some innocuous entertainment. It wasn't to be. Within a short while a news bulletin in large block letters covered the screen. A 10,000 lb SPACECRAFT HAS BEEN ORBITED BY THE SOVIET UNION. We later learned that the spacecraft weighed 10,015 lb and was in an orbit of 65° inclination, with a period of 91.25 minutes and apogee-perigee of 225 by 194 statute miles. If this large satellite was attitude-controlled and had a reasonably low-area-to-mass ratio (its frontal area, "facing the airstream" in the direction of motion, divided by the mass in pounds of the satellite) it would be up for a long time I thought. The announcement that a dummy cosmonaut was aboard was reason enough to dismiss my thoughts about long life, for this

---

10   Manned Space Stations Symposium, April 20-22, 1960, sponsored by IAS/NASA/RAND Corporation, Los Angeles, California.

was a test for a manned vehicle and an attempt to recover it was inevitable.

Further clarification from the Soviet Union's news agency TASS supplied additional data: there was a cabin of over 5,500 lb containing the dummy, life support systems, communications facilities, and various control panels. An instrument compartment weighed 3,250 lb, and various scientific experiments were also aboard. To that I mentally added biological specimens—mice, rabbits, guinea pigs, seeds, bacteria cultures, and the like. I never was able to confirm the presence of such specimens. A quick bit of arithmetic showed some 1,200 lb remaining for a retro-engine, propellants, and related controls. The cabin weight I presumed included parachutes for recovery of the cabin for, among other things, examination of the anthropomorphic dummy, which I was convinced had been instrumented to yield data on some of the effects of launching and recovery.

All of this speculation on results and recovery was washed away when on the 64th orbit—after four days of orbiting—an attempt was made to bring the craft down. The retro-engine operated at 2352 GMT on May 19 but some mishap in the attitude control system yielded an incorrect orientation during engine operation and instead of reentry the spacecraft went into a higher long-lived orbit. Moreover, the craft broke into at least eight pieces. Whether this was a deliberate act of destruction by the Soviets after the failure was never ascertained. The cabin portion was fairly intact for it continued to broadcast for eight days, life support conditions suitable for man were maintained that long according to Soviet reports.

On September 5, 1962, the main portion of the debris that was *Sputnik IV* descended over Wisconsin in the US in the area of

Chapter 2: Man in Space

Manitowoc. Several pieces were picked up and turned over to US scientists for examination.[11]

After the failure to retrieve the *Sputnik IV* spacecraft the Soviets took several months to review their gains and losses and did no launching in the interim. The review process paid off. On August 19, 1960, at 0844 GMT *Sputnik V* was launched from the Tyuratam complex. In the 10,143 lb vehicle were two very special passengers, Belka and Strelka (Little Squirrel and Little Arrow), a pair of female dogs. They "sailed" around in their 211 by 190 statute mile,[12] 65°, 90.72 minute orbit for 17-plus orbits, after which they were recovered by the Soviet Union at about 1045 GMT the next day. The recovery area was about 400 miles east, southeast of Moscow near the town of Saratov. The dogs, which had been observed via TV, were in excellent health and Strelka, at least, is known to have borne litters of healthy, normal pups sometime afterward. They are a famous pair of canines, the first animals, mammals at that, safely recovered after a space journey constituting a whole day in Earth orbit. The Russians were jubilant and rightly so.

There had been some early and fleeting misgivings when the dogs, observed via video immediately after reaching orbit, looked limp and unmoving. In fact, they were exhausted by their short but exciting launch into space and turned out to be merely resting. Before the flight, on the ground Belka had a pulse rate of 75 and a respiration rate of 24; for Strelka the values were 90 and 60 (evidently slightly excited). In the spacecraft during the launching Belka's pulse rate reached 150 and respiration 240 while Strelka's values were 160 and 125, respectively. After one orbit (90 minutes) the pulse rates were Belka 72, Strelka 65, and their respiration rates were 12 and 24, respectively, an obvious return to normalcy. These rates and their

---

11 *Aviation Week & Space Technology,* December 2, 1962, pg. 52.
12 From this point on I will abbreviate statute miles by sm.

changes do not appear very different from those of cosmonauts recognizing that excited dogs' respiration rates are basically somewhat higher than man's.

The vehicle was literally a flying biological and physical laboratory: aboard were 40 mice, two rats, insects, seeds, cultures of bacteria, cancer cells, human and rabbit skin samples, and for the physics experiments; thick-layer emulsions for cosmic ray study, instruments for X-ray and ultraviolet ray study of the Sun, and dosimeters for determination of radiation dose in the cabin. The Soviet craft represented efficiency—many of their space researchers were being satisfied in a single flight in addition to the preparation for manned flight that was in the offing.

In recovery procedures, the dogs and other animals in their container were catapulted out of the descending cabin (at an altitude of 23,000 ft) after the reentry process was completed. Reentry itself led to no more than 10 g, which evidently did not have any more than a momentary effect on the dogs. The actual retro-engine burn to initiate reentry commenced with orientation of the spacecraft rocket end forward in the direction of motion and pointed several degrees above the horizon. The engine operated for about 40 seconds during which time some 500 feet per second were removed from the spacecraft velocity. Both the retro-engine package (a liquid propellant engine) and the instrument compartment that carried the scientific packages for the physics studies were jettisoned. The semispherical cabin, it alone protected for reentry, continued to safely reenter the atmosphere. It aligned itself by virtue of its "uneven" weight distribution (i.e., it was aerodynamically stable) while roll jets kept it in a slow spin to "wash out" any remaining dynamic unbalance.

This great success by the Soviet Union not only whetted our intellectual appetites for more of the same but I and my colleagues also actually fretted as the days and weeks went by and no activity took place. Finally on December 1, 1960, at 0730 GMT another launching. Again, two dogs in an orbit. This time it was 117 by 159 sm, 65°, and with 88.6 minutes for the period. Pchelka (Little Bee) and Mushka (Little Fly), however, were not as fortunate as their two earlier companions. In attempting to retrieve these two dogs, the spacecraft returned on a trajectory much too steep for survival and the spacecraft was destroyed by the heat pulse.

Several things could have occurred; two that are possible are an excessive burn by the retro-engine thus taking off a large velocity increment and causing a steep, very hot reentry or perhaps the instrument compartment and/or retro-engine did not separate, which would cause the reentering craft to tumble. Since heat protection is only on a specific part of the recoverable cabin, tumbling would expose unprotected parts of the craft to huge temperatures with resultant incineration. Only a Soviet-written history of the event will tell us the exact nature of the malfunction that took place. Of course some experiments were conducted during the day that *Sputnik VI* orbited, but they were rather secondary considerations in view of the loss of the spacecraft and the two dogs. Nevertheless, for the first time, successful investigations of the solar X-rays were conducted; the solar corona was determined to have a temperature of 1.5 million degrees Celsius.

This great disappointment undoubtedly had a disheartening effect on the Soviet team and the resultant investigations as to what went wrong kept manned flight preparations out of the news for several months. In the interim the Soviets initiated their efforts in the interplanetary arena with an impressive commencement, about which I will have much to say in Chapter 7, "Soviet Reconnaissance Satellites: The Military World." Meanwhile, the workhorse booster of their space program was prepared for another flight at Tyuratam. The booster configuration, which was to surprise everyone who saw it in photo or on display at Le Bourget Aerodrome in France in 1967, was brought to the pad after being assembled and largely checked out while in a horizontal position. The latter, as any but the most casual observers would note, is quite different from the US technique of erecting the booster at the pad and performing all of the checkout there. (I refer, of course, to the days of Atlas and Titan, long before the era of Saturn rockets.)

The fourth precursor for manned flight was launched at 0630 GMT on March 9, 1961. The orbit parameters were similar to its predecessor; 155 by 114 sm, again at an inclination of 65° and having a period of 88.59 minutes. Aboard were a dog—Chernushka (Blackie)—and a zoo full of other animals; guinea pigs, mice, frogs, and other unnamed biological specimens (no doubt, an array of seeds, cultures, and the like). Unlike the previous craft this one, *Sputnik IX*, was aloft for only one orbit while the others were up for 17 orbits before recovery was attempted. I've often wondered if the first manned flight was originally intended for a whole day— 17 orbits—but had its plan changed after the problems with *Sputniks IV* and *VI*. That too must await revelation by some Soviet authority. This 10,362 lb spacecraft also had a "cosmonaut mannequin" aboard (an anthropomorphic dummy). The dog and other animals were evidently all observed using TV. Announcements by the Soviet Union were cautious and limited but clearly it was implied (some actually so stated) that a manned flight was in the offing.

I awoke on Saturday morning, March 25, 1961, to the news that once more the Soviet Union had orbited and recovered a dog named Zvezdochka (Little Star) along with a collection of animals and other specimens as on *Sputnik IX*. *Sputnik X* also carried an anthropomorphic dummy. The spacecraft weight once more was an impressive 10,351 lb; more than three times the weight of the US Mercury spacecraft in orbital configuration.

VOSTOK 1

At 0607 GMT on Wednesday, April 12, 1961, the dream of a millennium; no! the dream of many millennia came to life on a windswept, bleak semi-desert near an obscure town called Tyuratam, which is some 60 miles east of a larger, but to most Americans, just as obscure a city called Novo Kazalinsk on the eastern shore of the Aral Sea. The *Vostok* spacecraft, which gave its name to the booster that bore it toward the heavens, roared upward on its 800,000-plus pounds of thrust, spewing behind it a sea of yellow-white flame from 32 rocket nozzles. Twenty main nozzles and twelve control engine nozzles. Within 2 minutes the conical strapped-on engines plus tanks had expended their propellant and were jettisoned, the central sustainer continued to burn for another 3 minutes, finally it too was expended and jettisoned and the last stage, depicted in so many photos with it took the *Vostok* spacecraft into orbit. Sometime less than 15 minutes from liftoff Soviet Air Force Major Yuri Alekseyevich Gagarin was the first man from the planet Earth to look down upon it from Earth orbit. His reactions are familiar to all of us: "Beautiful, beautiful...I can see tilled farmland and forests in shades of brown and green, I see rivers and valleys...beautiful, beautiful." And history having been written, not even an ocean of tears will erase one iota...the Soviets were first again this time with a man in space.

Gagarin broadcast to the world on at least three frequencies; 9.019, 20.006, and 143.625 MHz. He may just as well have broadcast on every frequency in every land on Earth. Word spread like wildfire and both the public and their noisy governmental voices, the politicians were heard in the land. The prognostications were thick and unbounded; guesses on the shape and technological sophistication of Vostok gave the Soviet Union credit beyond their wildest dreams. The facts would not come out into the open for years. Gagarin landed near Krasny Kut in the Saratov region of European Russia near latitude 51° north and longitude 47° east. He quickly received a telephone call from Nikita Khrushchev; "You have made yourself immortal because you are the first man to penetrate into space." Gagarin replied, "Now let the other countries try to catch us."[13]

Gagarin's flight evidently awakened every wishful thinking anti-Soviet liar on the face of the Earth. Rumors of presumed earlier attempts to orbit a man appeared under many a byline well known in many a newspaper even better known. There were stories of disaster ranging from destruction of the booster on the pad to emotional derangement of the "astronaut" after his flight with every conceivable variation in between and some so unique as to be technologically impossible. When the new Astronomer Royal of Great Britain, Richard van der Riet Woolley, was told of the launching of *Sputnik I* after an airplane trip he said, "It's utter bilge,"[14] being one of those who, no doubt, thought that only the Western world could produce such an accomplishment (maybe he thought no one could, who knows?).

Because of my gentle nature I'll use his words to express my opinion of the committee of prevaricators. Hero of the Soviet Union Gagarin landed at 0755 GMT making his total flight time some 108 minutes of which some 70 minutes were spent in actual weightlessness. At 0725 GMT his retro-engine was ignited while he was in the area over equatorial Africa, at about 0735 GMT *Vostok 1* entered

---

[13] "Science: The Cruise of the Vostok," *TIME, April 21, 1961.*
[14] "Science: Utter Bilge?" *TIME, January 16, 1956.*

the atmosphere's denser portions, weightlessness was over. Within minutes reentry was complete, the drogue chute had activated followed by the main chute and the reentry cabin rode down to a soft landing. Gagarin stayed in the cabin unlike his compatriots to follow him into space in the remainder of the Vostok series.

Among the very significant information that we learned early was that the Soviets had chosen to keep the cabin atmosphere essentially the same as that of sea-level Earth. The variations from this during flight were of little concern, utterly negligible. The reasoning was as neat as could be asked for: if the atmosphere is kept the same as our everyday breathing mixture, then that will remove from special consideration the effects of some exotic atmosphere like pure oxygen and hence remove the burden of a whole range of problems. There are enough problems without unnecessarily adding to the list. The orbital parameters for *Vostok 1* were familiar; 187 by 109 sm, 64°54', and a period of 89.1 minutes.

The choice of perigee was notable also; should the retro-engines have failed to fire, the spacecraft would decay (i.e., reenter without the use of any artificial mechanisms), because of friction with the thin, but not negligible, remnants of the atmosphere at altitude, within 10 days. The spacecraft had adequate supplies for that length of time aboard; oxygen, water, food, and other needs. Some short time after the event, I estimated the size of the *Vostok*, its length and diameter, and calculated its decay time in the event of no retro-fire. I assumed a worst case, that is, with the nose of the spacecraft always pointing into the "airstream"; the positioning would account for the longest flight time under the given orbital conditions. My answer was under eight days. Years later when the real dimensions were released, I recalculated the decay time, it was about 7½ days.

Like several other astronauts in the US and one other in the Soviet Union, Gagarin did not live to see the accomplishment of the decade of the 1960s, the manned landing on the Moon by [Neil] Armstrong and [Buzz] Aldrin. He died in an airplane accident on March 27, 1968, just 18 days after his 34th birthday. The official announcement read, "The Central Committee of the CPSU (Communist Party of the Soviet Union), the Presidium of the Supreme Soviet of USSR and the Council of Ministers of USSR are deeply grieved to announce the death of Yuri Gagarin, the first cosmonaut, in an air crash."[15] Gagarin tragically died on March 27 as a result of a disaster during a training flight in an aircraft. Engineer-Colonel Vladimir Seryogin also perished in this air crash. Gagarin and Seryogin were buried at the Kremlin Wall in Red Square. While it is true that many cosmonauts and astronauts have gone on to greater achievements, Gagarin was man's first representative in Earth orbit, his niche in history is carved deeply and history will always remember his was the first flight.

Depending on the authority being quoted additional manned flights were to occur within days or weeks following Gagarin's flight. Rumors flew like snowflakes in the Arctic; they included speculations on multi-manned flight, winged-reusable spacecraft (shades of the shuttle now being built by both the Soviets and the US) and even remarks on manned lunar flight within the year (i.e., in early 1962). As the months passed the speculation hardly diminished.

---

15 "From the Past Pages of Dawn: 1968: Fifty Years Ago: Gagarin Dies in Air Crash," *Dawn, March 29, 2018.*

Vostok 2

On Sunday, August 6, 1961, at 0600 GMT, Major Gherman Stepanovich Titov was launched into orbit from Tyuratam. His 10,430 lb spacecraft was essentially identical (no two spacecraft for manned flight are really identical to this day; each flight leads to improvements in the next spacecraft scheduled to fly and so on downstream in time) to that of *Vostok 1*. However, *Vostok 2* was scheduled to make history again, for Titov stayed in orbit for more than 24 hours and despite the earlier speculation concerning more complex flights, the world stood wide-eyed in wonder.

Telemetry from Titov's ship was broadcast on 15.765 MHz, voice on 143.625 MHz and 20.006 MHz, while a beacon for tracking purposes was on 19.995 MHz (the latter frequency is used on most, if not all, Soviet reconnaissance spacecraft to this day). His orbit was fairly similar to that of *Vostok 1*; 159.7 by 110.6 sm, 64°58', and with a period of 88.458 minutes. The slight decrease in apogee was to be a characteristic of all *Vostok* flights. The interpretation of this publicly unnoticed alteration was a lowering of the energy of the orbit and hence implied a faster decay if it had to be done by nature rather than the retro-rockets. This also made the orbits more circular by a factor of two; the eccentricity of *Vostok 1*'s orbit was 0.009 while the other Vostoks were close to 0.004.[16]

Think of a camera in a satellite with good resolution, then look at Figure 1 (*Vostok 2* Earth traces) to get an idea of the coverage of

---

16 Appendix A gives a simple relationship for calculating the eccentricity of any Earth orbit.

the Earth's surface that is obtained in just one day in a close Earth orbit—Titov went around the Earth 17 times. Most reconnaissance satellites go around the Earth 200 or more times during their missions these days—think of the excellent and repeated coverage obtained that way. It makes one aware that no large structure above or below ground can be hidden from any country that has the capability to orbit a satellite of a few thousand pounds. If the below ground remark puzzles you, just recall that in building a large underground structure many trucks and their telltale tracks give away the game during construction.

Titov got quite ill for a time during his orbital sojourn and, for a while, this scared the hell out of the biomedical people—the stories of the difficulties of living in a weightless environment again reared their Medusa-like head. Titov, it turned out, was just one of those somewhat more sensitive persons insofar as his vestibular apparatus (balance sensors) was concerned. When he sat relatively quiet in his seat, the problem was alleviated. From time-to-time others would be similarly affected but only mildly so, never enough to have any lasting effect on a mission. Most cases would be barely discernible.

Titov's flight lasted 25 hours 18 minutes from liftoff to landing; he landed again in the Saratov area at latitude 51° north, longitude 48.7° east, at 0718 GMT on Monday, August 7, 1961.

The Soviets are much less restrained than their American counterparts, which fact is clearly indicated in Gagarin's radiogram to Titov during the course of flight. "Dear Gherman—I am wholeheartedly with you. I embrace you, my pal. I kiss you fervently. I follow your flight with great excitement. Confident in the successful accomplishment of your flight, which once more glorifies our great motherland, our Soviet people. Yours, Yuri Gagarin."[17]

---

17 *Pravda, August 7, 1961.*

Titov did not ride his spacecraft to the ground. Although it was never announced as such it appears that the *Vostok 1* landing was something of a jolt. As a consequence, Titov was ejected from his capsule after the reentry process was completed and the spacecraft had reached an altitude of 6500 meters (21,320 feet). The official release reads as follows:

Descent Timetable
(Greenwich Time)

1. At 0627 hours (on the morning of August 7, 1961) the ship's orientation system was activated by the automatic reentry mechanism.

2. At 0652 hours 20.4 seconds, the retro-rockets were fired automatically.

3. At 0711 hours the door of the ejection hatch was blown off automatically. (The spacecraft was at an altitude of 7000 meters, 23,000 feet, when this occurred, and the spacecraft was descending at a rate of 220 meters per second—722 feet per second at this time. Two seconds later at 6500 meters—21,320 feet the cosmonaut was ejected and his parachute opened.)[18]

4. At 0711 hours and 15 seconds the ship's landing system was automatically activated.

5. At 0718 hours the cosmonaut landed.

---

18 Several things should be noted here: A diagram accompanying the release shows the cosmonaut being ejected at 6500 meters but nothing in the text specifically states this, so that I inserted it parenthetically. The rate of descent is also given by the Soviets in the diagram as is the two-second delay for catapulting (automatically).

The remainder of 1961 again brought one of those large gaps in the Soviet space program. There were no other flights manned or otherwise until March 1962 when the Cosmos series was initiated and the announcement, "flights from different cosmodromes will be made," was issued by the Soviet Union. Finally, things became interesting once more when *Cosmos 4* was launched on April 26, 1962, and recovered three days later after 46 orbits and *Cosmos 7* was launched on July 28, 1962, and recovered four days later. There were a few curious characteristics about both of these satellites that aroused my interest beyond what I would call normal for that era. *Cosmos 4* was recovered in the Saratov area. *Cosmos 7* was the first satellite recovered in a completely different area from all of the earlier recoveries. It was recovered near the city of Karaganda, a mining center in Kazakhstan (Kazakh Soviet Socialist Republic) located at about latitude 50° north and longitude 73° east. Karaganda was to be the recovery center for almost all Soviet recoverable satellites in the future. For neither one of these satellites was the launching time announced—and for me that was disturbing, for how could I then determine the total mission time, the number of orbits, and the launch base, and tie all these items together?

I decided that if launch times were to be kept secret, there was a challenge to me personally to exercise my ingenuity so that this would be secret no longer. If I did this correctly then there would be a way of checking whatever techniques I developed through information released by the Soviets themselves. Suffice to say now, I succeeded, and I'll explain the method a bit later. The complexity is minimal and will be easily understood; but now back to our manned spaceflights.

I recognized the fact after the next flights (though had I collected telemetry data like, say, the Space Institute in Bochum, West Germany) I am reasonably certain that the basis for the *Cosmos 4* and 7 flights could have been ascertained before the flights of *Vostok 3* and *4* took place. The Cosmos satellites were precursors for manned flights; *Cosmos 7* duplicated the flight for *Vostok 3* and *Cosmos 4* was the "test" vehicle for *Vostok 4*. The precursor flight vas to appear for a long time in the Soviet manned spaceflight program.

According to I. Zaitsev[19] *Cosmos 4* also tested the procedures of taking photographs of cloud cover using video equipment while *Cosmos 7* had equipment for getting a reading on the radiation environment in space. Both of these satellites had orbits with higher apogee and perigee than the Vostoks, so that it appears that equipment aboard was being tested together with reentry and other procedures rather than any particular facet of the orbit parameters

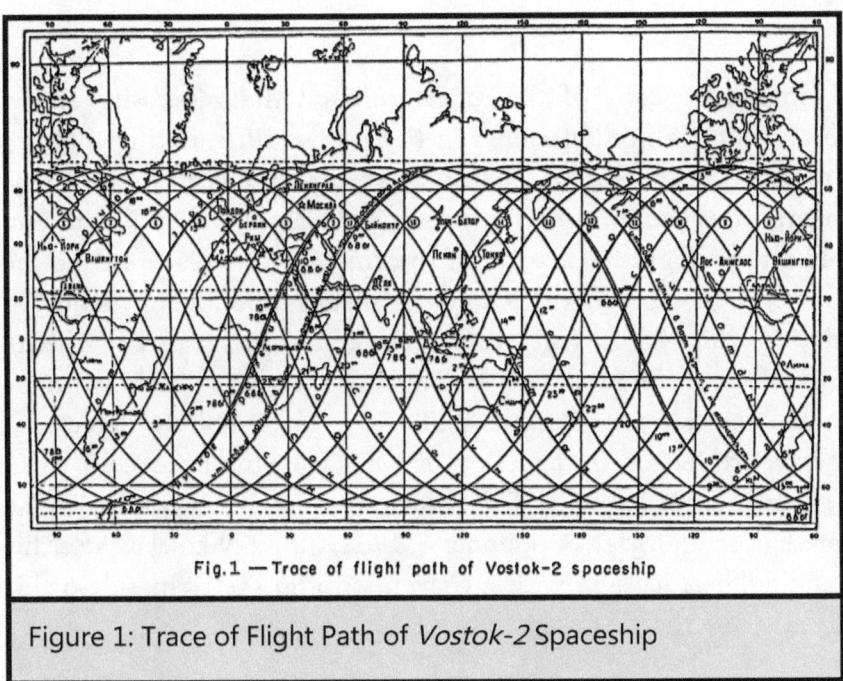

Figure 1: Trace of Flight Path of *Vostok-2* Spaceship

---

19 "The Satellites," I. Zaitsev, *COSMOS*, Moscow, Znanie Press, 1970.

themselves. There is no question about the ability to duplicate the orbit itself, it can be done with relative ease.

While getting dressed on Saturday morning, August 11, 1962, as is the habit of millions of us, I turned on the radio to pick up the news. It was delightful to hear that still another cosmonaut Major Andrandrian Grigor'evich Nikolayev was in orbit onboard *Vostok 3*. The "safe" orbit was used again; 141.4 by 109.8 sm, 64°59', with a period of 88.2 minutes. Of course, I was certain that this flight would exceed the Titov flight but the question was by how much; I guessed the flight would last three days. Nikolayev was a member of the Chkalov flying club as had been Gagarin and Titov and they offered a report on the flight, which had begun at 0829 GMT. Most of the data they released was made known through the *FAI Report* (Federation Aeronautique Internationale) on the flight.

I obtained a copy of the report through the generosity of the FAI, which is headquartered in Paris. It is worth noting that the information therein was fairly substantial, though not nearly as plentiful as the information released by the US, both before and after their flights. All sorts of information was discussed in the Soviet press regarding the weightlessness studies being conducted during this flight (the Titov syndrome still worried them). Later photos showed that among the instrumentation worn by Nikolayev were nystagmus indicators (eye motion detectors), which are related to detection of balance disturbances and EEG (electroencephalograph) for brainwave studies. A photo in the press from Thursday, September 6, 1962, shows Nikolayev after his landing, bearded and with a shaven strip on the right side of his head where the EEG sensor had been attached.[20]

---

20   See the *San Jose Mercury* of that date, for example.

On the following morning, Lieutenant Colonel Pavel Romanovich Popovich followed his compatriot into orbit, lifting off from Tyuratam at 0802 GMT on Sunday August 12, 1962. The time was such that Popovich in *Vostok 4* was to make a very close approach to *Vostok 3* piloted by Nikolayev. The uproar that ensued over this pseudo-rendezvous attempt lasted for months. It reached the highest levels of the US government and was splashed across the front page of innumerable newspapers both in the US and abroad. The essence of the question was, did the Soviets attempt to rendezvous, and did they succeed? The arguments started while the cosmonauts were still in space and broadcasting their personal and their country's greetings to all the world.

Vostok 4 used the same frequencies as *Vostok 3* except for the tracking beacon, which was now 19.990 MHz. Ya Sokol (I am Falcon) broadcast Nikolayev; Ya Berkut (I am Eagle) radioed Popovich from their orbital eyries. The orbit parameters for Popovich were 145.9 by 110.54 sm, 64°57', for the inclination and a period of 88.3 minutes. After looking at the data, I decided to get into the act also and Figure 2 is exactly as the original calculation I made and passed around to my colleagues. All of the data came from *Pravda* editions of August 12 and 13, 1962.

Figure 2 shows that the two spacecraft were in orbital planes only 2 minutes apart at apogee, the orbital heights differ by only 4.5 miles, while at perigee they differ by only 0.74 miles. If they were at apogee simultaneously the slant distance between them, shown in Figure 2, puts them 5.1 miles apart; if simultaneously at perigee, then only 2.76 miles apart. The question I couldn't answer was where were they in their respective orbits early in the flights.

Sir Bernard Lovell is quoted in *Aviation Week & Space Technology* of August 27, 1962, as saying that his large antenna at Jodrell Bank (near Manchester, England) could not resolve two objects early in

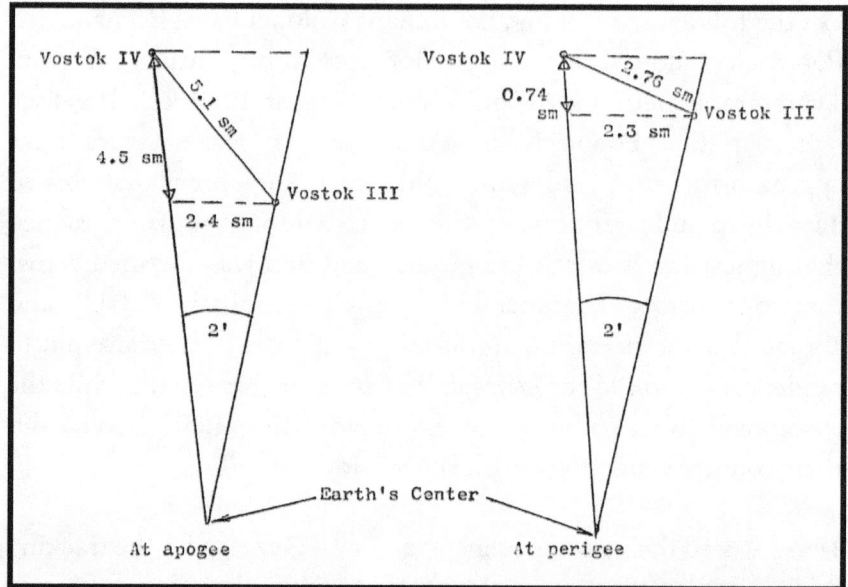

Figure 2. The Flight Paths of *Vostok 3* and *4*.

The diagram defines the trapazoidal slot that the two Vostok spacecraft occupied during their orbital sojourn. Since they passed each other at sometime during their traverse, the bounds of the distance between them is shown.

Pertinent Data	Vostok 3	Vostok 4
Orbital Period	88.2 min	88.3 min
Inclination	64° 59'	64° 57'
Apogee	141.42 sm	145.8 sm
Perigee	109.8 sm	110.54 sm

the flight (but presumably could "hear the broadcasts") so that the separation could only be a few miles at most. Since Lovell made that statement, I was satisfied that they were close, indeed. The same sort of numbers that I calculated also were announced by the Russians.

But they said, contrary to opinion stated in the press, that they did not attempt a rendezvous. The flights from that aspect were an investigation into procedures for some future rendezvous—evidently a practical try at seeing how close they could get a second booster off to calculated time for a real rendezvous. The answer was obvious, damn close!

In broadcasting, Nikolayev said that he could see Popovich's craft as "a little moon" a few miles distant. When the FAI reports were made available, for me at least, the question was settled conclusively. The reports clearly show that there remained a 0.01 kilometer per second (33 feet per second) velocity difference between the craft throughout their flight.

The arguments were not nearly finished. Deciding that the information I had put out in the memo described in Figure 2 was useful and interesting, I wrote a letter to the magazine *Missiles and Rockets* giving all the details—and it was promptly published in the issue of September 3, 1962. Wishing to avoid the appearance of having an exact answer in the mathematical sense, I made a remark in the letter that, "most calculations made are necessarily somewhat inexact," because of the relatively rapid changes in perigee, apogee, and period for such low orbits. In the October 1 issue of the magazine there appeared a letter raising Cain with my calculations and pointing out that, in considering the different launch times for the two Vostoks, I had neglected the rotation rate of the Earth and that therefore, the spacecraft must have been several degrees apart when *Vostok 4* was launched; this means 70 miles apart for each degree. And if several is as few as three, then the spacecraft were separated by 210 miles. The writer was a science editor from Holland. The problem was that the poor chap really didn't understand orbital mechanics; didn't recognize changes made by the Earth's oblateness, which causes orbital planes to precess to the west (regression of the nodes); and didn't know how to calculate the liftoff time for the rendezvous of a second spacecraft with one already in orbit. I wrote a second letter explaining in detail just what had to be done in a proper calculation, it was promptly published. There was no further rebuttal, end incident.[21]

The group flight lasted until August 15; Nikolayev landed at 0652 GMT near the town of Karkaralinsk at latitude 48°2' north and longitude 75°45' east, a few miles southwest of Karaganda; while Popovich landed 7 minutes later at Atasu a few miles south east of Karaganda at latitude 48°10' north and longitude 71°51' east. The displacement on the ground of the landings was due to different retro times, probably slightly different spacecraft orientations, the rotation of the Earth between retro and landing, and the actual variation from the calculated times and positions for retro and orientation. The Soviet reports on the flights make clear the fact that they consider the cosmonauts' flights over at the time they are ejected from their respective spacecraft.

Thus, Nikolayev's flight lasted 94 hours and 10 minutes while Popovich's flight was 70 hours and 44 minutes. After ejection, the pilots reached the ground within 12 minutes, descending at a rate of about 30 feet per second from their catapult point 21,320 feet above the ground. I noticed that there is some difference in descent time between Nikolayev/Popovich and Titov that is unexplained: according to the Soviet records, Titov came down after ejection in only 7 minutes but for the same procedure, the *Vostok 3* and *4* cosmonauts took 12 minutes to reach the ground. This might be easily explained by assuming that the Titov text was prepared beforehand and, in practice, some change had occurred, which was not picked up in the Chkalov flying club report. Of course, I'm guessing and the discrepancy remains.

---

21   See Appendix B for letter.

The length of the flights staggered the imagination of the public and the Soviet Union made the most of their advantage in large booster capability. That advantage was lost when the Saturn class boosters came into use in the United States but the perseverance that has always marked the Soviet program is still with them as strong as ever; same could hardly be said of the US program. The US Congress, ever the parochial house, simply hasn't got the foresight to finance long-term activity for the NASA missions. Instead, NASA gets fed piecemeal and its programs are subject to the whims of [Democratic Senator William] Proxmires and [Vice-President Walter] Mondales. If not for the superb leadership of President John F. Kennedy there never would have been a callback from the Moon, "Tranquility base here. The *Eagle* has landed." The present [1975] leadership has assured us that the US will once again fall far behind the Soviets and this time it will be much more difficult to catch up.

* * *

For the next year the Soviets had a frustrating time in trying to get their interplanetary program in motion and had a long string of failures to fly by Mars and Venus. On the other hand, their efforts in the reconnaissance arena were much better. They had seven spacecraft orbited and recovered, one staying in space for over 236 hours, *Cosmos 16*. However, these were all part of unmanned activity about which little was said except that they were placed in orbit. At least two attempts to reach the Moon with unmanned soft landers also failed to reach their targets.

There remains to this day a considerable question about whether or not *Vostok 5* and *6* were supposed to fly a group flight in the manner that *Vostok 3* and *4* did. Contrary to numerous remarks in the press and an announcement by the Soviet Union that the two ships passed within some four or five miles of each other they did not repeat the performance of Nikolayev and Popovich. When the *Vostok 5* and *6* spacecraft passed each other, their relative speed was 12,000 feet per second while in the earlier case a small difference of 33 feet per second was kept throughout the flight. This difference is also pointed out by the fact that the two spacecraft were launched at a time difference not commensurate with, congruent, or even closely similar orbits. The two orbit planes were 26°45' apart, possibly a bit more. At about 0945 GMT on June 16, 1963, the spacecraft made their "closest approach," the four or five miles referred to above. The two pilots may have had radio contact but they did not see each visually. This "flash-by" took place over the northeastern sector of Siberia.

* * *

Lieutenant Colonel Valeri Fedorovich Bykovskii was launched into space from Tyuratam at 1159 GMT on Friday, June 14, 1963 on *Vostok 5*. His orbit of 138 by 108.7 sm, 64°58', with a period of 88.27 minutes was incrementally lower than those of his predecessors. His tracking beacon was on 19,948 MHz while the other frequencies were the familiar ones. His flight was almost five days in length, 118 hours and 57 minutes to the point of being catapulted from his spacecraft. Bykovskii landed in the Karaganda area, 335 miles northwest of that city near a town called Peski at latitude 53°24' north and longitude 67°37' east.

The launching of Valentina Vladimirovna Tereshkova at once delighted the world and brought some inane remarks from some persons associated with the US space program. Some men can't stand the thought of competition from a woman. Alas for them, Miss Tereshkova had the rank of Junior Lieutenant in the Soviet Armed Forces and, prior to her becoming a member of the world's most exclusive club, appears to have had little aircraft flight training. Clearly, she is to be admired for her accomplishment. By the time she had completed her flight she had orbited the Earth more times than all of the Mercury astronauts in their total program: 48 orbits to 35 for the Americans.

Tereshkova was launched in *Vostok 6* at 0930 GMT on Sunday, June 16, 1963, into an orbit of 144.5 by 113.5 sm, at 65°, and with a period of 88.3 minutes. At about 0810 GMT on June 19, Tereshkova was catapulted from her spacecraft and she landed about 385 miles northwest of Karaganda near a town called Zav'yalovo close to latitude 53°16' north and longitude 80°28' east. Her total time in flight was 70 hours 41 minutes to the catapult point. It is interesting to note that Miss Tereshkova's flight was intended to be a one-day affair and was to be extended only if everything was completely satisfactory in flight. The record speaks for itself.

A particularly happy event followed Miss Tereshkova's flight within five months. She was married to Cosmonaut Andrian Nikolayev. To quote from the report of the *New York Times* of Monday, November 4, 1963; "Moscow, November 3—The world's first spacewoman, Captain Valentina V. Tereshkova, was married today to a fellow astronaut, Major Andrian G. Nikolayev, in an emotional civil ceremony at a Moscow wedding palace. At a four-hour reception for 300 guests, Premier Khrushchev led off a long series of toasts. The reception, in a government guest house usually reserved for state visitors." Mrs. Tereshkova-Nikolayev has since had at least one child, however, one should not expect her to be necessarily removed from flying status permanently.

All of the Vostok flights were intended to investigate the capabilities of man in the weightless environment and, to that end, all of the spacecraft carried numerous experimental animals in addition to the cosmonaut-pilots. An array of experiments was conducted to examine the radiation environment including the Van Allen Belt as well as cosmic rays from the Sun and from beyond the solar system. Results from all of these were obtained in due course; an endless stream of technical papers have been published. Although none of them offered restrictions on the length of spaceflight, per se, because of the exotic combined environment, the Bykovskii flight of five days was to stand for a long time for the Soviets despite US flights well beyond that period of time.

It has been characteristic of spaceflight that although each of the major space powers has announced many results, still each prefers to find answers for itself whatever the question is, but particularly those concerning man. It should also be observed here that the Soviets have released much data in the experimental areas of spaceflight that do not have military implications. When the latter is true then there are simply gaps in the information. The Soviets have not rearranged the facts in any of their findings; when something is to be hidden it simply isn't discussed. On the other hand, when prestige is to be gained then all the knowledge collected is made available and quite readily so.

As is the custom everywhere, new data brings a preliminary sort of releases followed up by detailed, refined, and complete results. I have seen what must be thousands of their papers and have read some hundreds of them. I do not read Russian [Cyrillic script language], but I shall be forever indebted to my friend and colleague Morris D. Friedman who has translated all of the Soviet papers, but a few, that I have read. His diligent searches for me for some data necessary to answer some question or to fill a gap in information already at hand are such that I could not hardly begin to write this book without his aid. There are many good translators in the United States but few who have the insight and ability to detect a subtle nuance that he has. Moreover, he is a mathematician, and thus being on the side of science to begin with, translates scientific data with a fine advantage over most other would-be Russian students.

* * *

On October 6, 1964, the Soviets launched a spacecraft that they identified as *Cosmos 47* into the usual 65° orbit but unlike their reconnaissance satellites this one had an apogee of 256.6 and a perigee of 110 sm; the orbital period was 90 minutes. Several things were puzzling: the perigee was that used for manned flights, the apogee was not; neither was it low enough for the reconnaissance birds. To add to the mystery, it stayed in orbit only for one day, 16 orbits, 24 hours and 17 minutes. It was evident enough, mystery or not, that here was a precursor for something, but what?

The puzzle was not long in being resolved: on Monday morning at 0730 GMT, October 12, *Voskhod 1* was launched with not two but three cosmonauts aboard. The 11,728 lb spacecraft carried Engineer-Colonel Vladimir Mikhailovich Komarov, Scientific Officer Candidate of Engineering Sciences Konstantin Petrovich Feoktistov, and Medical Doctor Boris Borisovich Egorov into an orbit of 255 by 110 sm, at 65°, and with a period of 90.1 minutes. *Voskhod* (Dawn) landed the next morning, October 7, near the city of Kostanay, almost directly north of Tyuratam at latitude 53.2° north, longitude 64.2° east. The flight time was 24 hours 17 minutes!

Because of the short orbit time, a rash of rumors from the usual know-nothings spread like a prairie fire. Typical was a comment from *Die Tat*, a Zurich newspaper, which had a story to the effect that faulty ignition in a rocket stage had put the spacecraft in an incorrect orbit and hence the early descent, etc. Pure fertilizer. The matching of the *Cosmos 47* orbital time with that of *Voskhod 1* was hardly a coincidence. But a much greater surprise was revealed—the cosmonauts wore no spacesuits; there was no catapult mechanism; and the retro-engine had a backup engine that could be used for other maneuvers while still in orbit, if this alternative was desired. Additionally,

- The spacecraft was announced as stable for water landings and had touchdown rockets for the landing process (such rockets fire when the spacecraft is about 10 feet above the ground thus assuring a very soft landing, indeed).
- The environmental control system was improved to the extent that the oxygen content could now be varied on demand (the Vostok system was not under the control of the pilot).
- New orientation systems as well as new video and radio systems were aboard.

The spacecraft's velocity vector could be ascertained by the flow of ions in the wake of the spacecraft in orbit. Not only had the sophistication of the spacecraft increased by a quantum jump but also the booster had been upgraded to 1,000,000 lb of thrust. At this stage of spaceflight, I was most impressed with the confidence that the Soviets had shown in this first flight of a new vehicle—in that the crew wore no spacesuits and rode the bird all the way to the ground.

As expected, the doctor was along for a firsthand look at the weightlessness problem and to study of the intricacies of the vestibular apparatus of his fellow crewmembers. It strikes me as comforting to have one's own MD aboard in the event of a medical problem. There were none. The flight was relatively uneventful in the sense that any untoward occurrences must have been minor, for the crew looked fine in photographs taken within a few hours of their landing. I thought to myself at the time: It was 3½ years from *Sputnik I* to *Vostok 1* and 3½ years from *Vostok 1* to *Voskhod 1;* will it be just three-and-one-half more years to the Moon for the Soviet space farers? It was to be a longer interval than I had imagined.

On December 15, 1964, an address on Soviet bioastronautics was given before the members of the National Space Club in Washington, DC, by Dr. Eugene B. Konecci, who at that time was a member of the National Aeronautics and Space Council, a US government advisory body. During the course of his lecture Dr. Konecci made the following points, which are quoted here from his paper:[22]

1. Early extra-vehicular operation, then, was a key to the selection of the first American life support system. We are presently prepared to go extra-vehicular in the Gemini program. It appears doubtful at this time that the Soviets would attempt the same maneuver. There is no indication that the Vostok or Voskhod have airlocks.

2. Undoubtedly, in the early embryonic stages of manned flight that we are now experiencing, this extra-vehicular requirement may not be readily apparent or appreciated. I can assure you that the Soviet bioastronautics scientists are aware of it and are anxiously awaiting our Gemini experiments.

3. At this time, it then appears that the United States is ahead in the manned extra-vehicular area, but not far ahead.

These statements were made at the bare beginnings of the flight portion of the Gemini program, before manned flight was to

---

22 No reference given in original document for this quoted text.

commence. The first manned Gemini flight was scheduled for late March 1965. At 0659 GMT on March 18, 1965, *Voskhod 2* was launched from Tyuratam into an orbit of 307.6 by 107.5 sm, at 65°, with a period of 90.9 minutes. There were two cosmonauts aboard; Colonel Pavel Ivanovich Belyayev and Lieutenant Colonel Aleksei Arkhipovich Leonov. I heard all this as I was dressing and preparing to go to my office that Thursday morning. As I listened to the radio in my bedroom I wondered why the radio correspondent sounded so excited when he said that the two cosmonauts were talking to each other. I thought what in hell is so unusual about that?! At that point I switched off that radio, walked to the dining nook, turned another radio on, sat down and proceeded to enjoy my icy-cold, tart grapefruit. In the midst of a mouthful of grapefruit, I heard the reporter state that one cosmonaut was OUTSIDE OF HIS SHIP. My wife claims to this day that I jumped clear out of my chair without changing the sitting position. I knew now what the excitement was all about. I recalled Konecci's remarks and damn near choked with laughter; what an amusing turn of events.

Turn now to the Soviet command posts in Tyuratam and in Moscow, and they, listening to the flight's progress at 0730 GMT when *Voskhod 2* was over the Black Sea, heard the anxiously awaited signal, "Dawn! Dawn! This is Diamond (Belyaev), Man's walked out into space. He's afloat!"

Leonov had gone into space through the use of an extendable airlock. The airlock was extended and checked. Leonov left his cabin through a hatch, which was then closed. The airlock was depressurized, Leonov checked his systems—communications, life support, power-opened the airlock and cautiously crawled out and performed a number of assigned tasks. All of this was recorded on tape; he was watched by the ground stations in real-time live video transmission. The total time spent in the environment of space was about 20 minutes equally divided between time in the unpressurized airlock and the exercises in space outside the ship. By the time his extravehicular activity (EVA) excursion ended, the *Voskhod* was over Siberia above the Yenisei River, which flows from the southwestern tip of Lake Baikal to the Arctic. The airlock was then jettisoned.

All the time Leonov was outside he was carefully monitored by Belyayev in the event something untoward occurred and aid was needed. None was. Because the automatic guidance system malfunctioned, it was necessary for the spacecraft to be brought down by manual control. which, it appears, led to some inaccuracies in the landing. As a result of the malfunction, the cosmonauts received new instructions for their landing, completed one more orbit than originally called for in the flight plan, and landed near the northern Soviet city of Perm at about latitude 58.50 north and longitude 57.5° east. This was considerably farther north than any landings had been made previously. In fact, there are some tales that the cosmonauts had to retreat to their ship after landing and getting out because of the wolves that threatened them. Be that as it may, helicopters brought them out the next day. The 26-hour two-minute flight had made history, once again, signed with the seal of the Soviet Union.

In the March 19, 1965, edition of *Pravda*, there appeared on the front page, closely centered on that page, a paragraph of fewer than 10 lines; its words in effect stated that Dr. Konecci was not correct, the Soviet Union would not wait for information on EVA from the American Gemini program.

Aleksei Leonov had made a reputation in the world of art, and exhibits of his paintings can be found in the Soviet Union and in the book, *The Stars Are Awaiting Us*, written and illustrated together with Andrei Sokolov [published in 1967]. The fates were not kind to Belyayev; he died on January 10, 1970, of complications arising from an operation for stomach ulcers.

3

WHEN ARE THEY LAUNCHED?

With some rather few exceptions the launching times of Soviet spacecraft are not announced either beforehand or after the fact. When a manned launching occurs and the spacecraft is successfully in orbit, the launching time will be released by the Soviets. In the event of an important launch in the vistas of science like those to the planets and sometimes, but not always, for those to the Moon, the time will be given after the vehicle is successfully on its way. A spacecraft in any of these unmanned areas, which reaches Earth orbit but does not succeed in leaving this orbit and is supposed to for its mission, is either ignored (as occurred in the beginnings of space activity) or, in a slightly more realistic manner, is just called the next successive Cosmos and the launching time is not announced.

On occasion the spacecraft statistics will be withheld until it is evident that the entire mission has been a success. The United States authorities that monitor Soviet launchings are always under a cloud of practically pathological secrecy, despite the continual references to these activities in really reliable newspapers like the *New York*

Times and magazines like *Aviation Week & Space Technology*. Any questions to these authorities or their agencies result in a blank stare, a thunderous silence, or some inane remark but never the facts. The excuse is inevitably the same one repeated on all such occasions: if we reveal such information, we will be revealing our capabilities. On the other hand, one can obtain such information from releases in other countries, Great Britian for instance.

What all this amounts to is that, for the most part, eventually the information can be obtained. The Soviets, of course, know when they launch but the American public is kept largely in the dark, about this and about many aspects of Soviet space activity, thus suppressing knowledge and making some aspects of space activity unnecessarily cloudy or even mysterious. This whole approach is in vain for there is a technique of determining the launch time for almost all Soviet launches to date. I say almost all because there have been some small but annoying gaps in the information needed to find launch times.

It should be noted that orbital parameters are given in almost every launching so that it is the launching time alone that is the major parameter so frequently unknown because it is unannounced. Sources for information are the Soviet newspapers *Pravda*, *Izvestia*, and *Krasnaya Zvezda*, with *Pravda* usually carrying the most complete analysis of any given space event. Even results of scientific experiments conducted aboard spacecraft appear in these newspapers much as one of our US scientists would give a paper on some presumably new topic at a meeting of some technical society. Results for the Soviet experiments are published in technical journals in the Soviet Union as well as abroad in the same manner as in any other country.

Calculating Orbits

But back to the launch times. The orbit of any satellite can be ascertained by using radar equipment to track the satellite by either listening to whatever broadcasts are being made from the bird, by listening to its beacon by which the country of origin will track it, or by skin-tracking, which means bouncing radar signals off the satellite and recording the roundtrip time for the signals. The latter is a method of last resort since it takes much more power than the other methods. The essentials are the calculation of the velocity of the spacecraft and its distance from the surface of the Earth (and hence its distance from the center of the planet).

Accurate knowledge of these two parameters permit the values to be inserted into the equation for an ellipse and into its auxiliary equations. If this is all set up in a computer program, which is, indeed, the method used, the apogee, perigee, period, and inclination to the equator are determined. The more samples of the radius and velocity that are obtained the more refined is the result for the orbit. Such samples have to be taken at fairly frequent intervals since the remnants of atmosphere at satellite altitudes will cause changes in the orbit of any particular satellite from day to day.

The actual changes depend on the altitude; the lower the orbit the greater the change; the shape and size of the satellite—the larger the satellite for any fixed weight (on Earth) the greater the change; the condition of solar activity—an active Sun with many flares leads to a higher density of the Earth's atmosphere at any given height and thus an increase in the decay (orbit-radius decrease) for a satellite. There are also interactions with the Earth's magnetic field and at higher altitudes pressure from the Sun's light. At very high altitudes—say of the order of 30,000 sm—some satellites are greatly affected by the influence of the Moon. At the lower altitudes, say below 400 sm, the major influence is aerodynamic. Somewhere above 600 sm the influence of Sunlight is important enough to consider when trying to determine what the orbit life of a spacecraft will be in pre-mission calculations before launching. Of course, radar observations really don't give a hang about all of these separate influences; one listens, as it were, with the equipment and the connected computer prints out the resulting orbit.

All of this information is used to print out a series of bulletins upon which the characteristics of the orbit are given along with the time and angular location of the ascending node for each orbit pass of each satellite. The ascending node is the longitudinal location on the equator of the south-north path of the orbit trace on the Earth's surface. The Earth is not a homogeneous sphere, it is oblate and rotating, and, therefore, has a greater diameter at the equator so that a satellite orbit will "twist" to the west. This fact together with the Earth's rotating under the orbit makes the orbit appear to have a continual westward movement.

It should be recognized that due to the nature of the Earth's gravitational field all satellite orbits, at any instant, are in a plane passing through the Earth's center in the same manner as a plane containing any great circle on the Earth's surface. All of the planets of the solar system yield orbits under the same laws. The governing laws are derived from the mathematics set up for these phenomena by Sir Isaac Newton (1642-1727) and by Johannes Kepler (1571-1630).

This total movement per orbit is defined for us by the ascending node. The time of the ascending node is Greenwich Mean Time

(GMT), the time at longitude 0°. This time is measured on the 24-hour clock so that twelve o'clock noon is read as 1200 GMT and twelve o'clock midnight is read as 2400 GMT. The first two digits are for the hour and the second two are indicative of minutes; for 1304.33, read 13 hours, 4 and 33 hundredths minutes. The longitude in the bulletins is always stated as west longitude, starting at the zero meridian and moving westward around the Earth. Thus, a particular orbit location will be given as 24 0624.55 237.67—read as orbit 24 at 6 hours, 24 and 55 hundredths minutes, at longitude 237.67° west.

The method I use for obtaining the time is dependent on the fact that most Earth orbits are close to circular and are within a few hundred miles of the surface. History has shown that satellites launched from a particular base at a particular orbital inclination will have an initial ascending node that varies very little from launch to launch. The initial node, in space jargon (and in astronomy) is called the zeroth ascending node.[23]

In order to have a basis for calculating launches, regardless of launch base, all are measured from the zeroth node; that is, the calculations are made at some very early time in the satellite's lifetime and the orbit is calculated backward until the first south-north pass (the zeroth node) over the equator is reached. After the orbit calculation passes through the launching base the portion of orbit between the launch base and the equator is calculated— although it is FICTITIOUS. A glance at Figure 3 will reveal at once the truth of the latter statement. However, that segment of the orbit trace is the basis for my time calculations.

For any launch of a given inclination within the altitude restrictions previously noted, the time of launch is the time of zeroth node to which is added the incremental time from that node to the launch

---

[23] Recall that the launching countries have launch bases where it is convenient to them and, therefore, at different latitudes.

base. To determine the increment for that segment, I use spherical trigonometry—which I learned in high school—to find the angular length of the segment. The formula given below can be found in any mathematics or engineering handbook besides being available wherever trigonometry is taught. That angular length in degrees divided by 360° and then multiplied by the satellite period gives, to reasonable accuracy, the time to travel the arc in question.

Now, it is true that there are errors in this method. For example, the satellite being some hundreds of miles above the Earth, and its orbit being elliptical instead of circular will cause a variation in the true segment length for each satellite orbit. Since the orbit itself rotates (the line connecting apogee and perigee—called the line of apsides—actually rotates) the segment length changes. No matter, these are negligible for our purposes. Any spacecraft that goes into low Earth orbit or any lunar or interplanetary spacecraft that leaves some rocket case or other debris in a parking orbit around the Earth for which a bulletin is issued leaves an ironclad "calling card" from which the original launch time can be calculated.

The question arises, how good are these derived times? There is an incontrovertible technique for comparison of the times I derive with actual launch times; those launch times announced by the Soviets, whatever the mission. I have simply used my technique and compared the results with their times on a regular basis in order to test the method as frequently as possible. Agreement is not just good, it's amazing. In some cases, because of orbit peculiarities, my derived times for liftoff have been off as much as 5 minutes, a very bad result. Most results are better than 1 minute and often the error is down to seconds. I can hardly count the times, there are so many, that no differences existed at all. The use of these liftoff times is manifold; the reader can be sure that the liftoff times given in Table IX for all launchings are quite authoritative.

Table II contains a set of numbers derived for the three Soviet launch bases and for each of the several inclinations used at each of them.

It frequently happens that there is a failure, deliberate or otherwise, to issue a first bulletin on some satellites. Thus, one is faced with having no zeroth node handy from which to calculate a launch time. In this eventuality, I subtract the two earliest consecutive orbits, which gives the satellite period, and multiply this value by the earliest orbit number given. Then this number of resulting hours and minutes is subtracted from the same earliest orbit time, which then reveals the time of zeroth node. The same procedure applies to finding the longitudinal location of its node. An example of this is given below for a particular satellite. A negative result in finding the location is resolved by subtracting that number from 360 to find the location using the west longitude only method. A time calculated that is greater hours and minutes than the time given for the provided early orbit is resolved as follows:

a.) The stated (given) orbit time must always be greater in both hours digits and minutes digits than the derived time to be subtracted from it. The derived time, if over 24 hours, must have all multiples of 24 hours subtracted out first before using further.[24] If the minutes digits are greater in the derived time, the stated early orbit must be changed so as to reduce it by 1 hour and add 60 minutes to the minutes digits for that orbit.

b.) If in calculating for the location, the number of degrees derived is greater than 360, all multiples of 360 must first be subtracted from it before using with the bulletin-stated values. IF THE READER FINDS SOME OF THIS

---

24 Each 24 hours subtracted represents one day back in time.

MATHEMATICS COMPLEX OR UNINTERESTING, I SUGGEST THAT YOU GIVE UP ON THIS AND GO ON TO THE NEXT CHAPTER. MOREOVER, YOU MAY READ THE TABLES FOR THE DERIVED LIFTOFF TIMES AND IGNORE ALL THESE NUMBERS.

For my colleagues who may have some interest the following example may be of same aid.

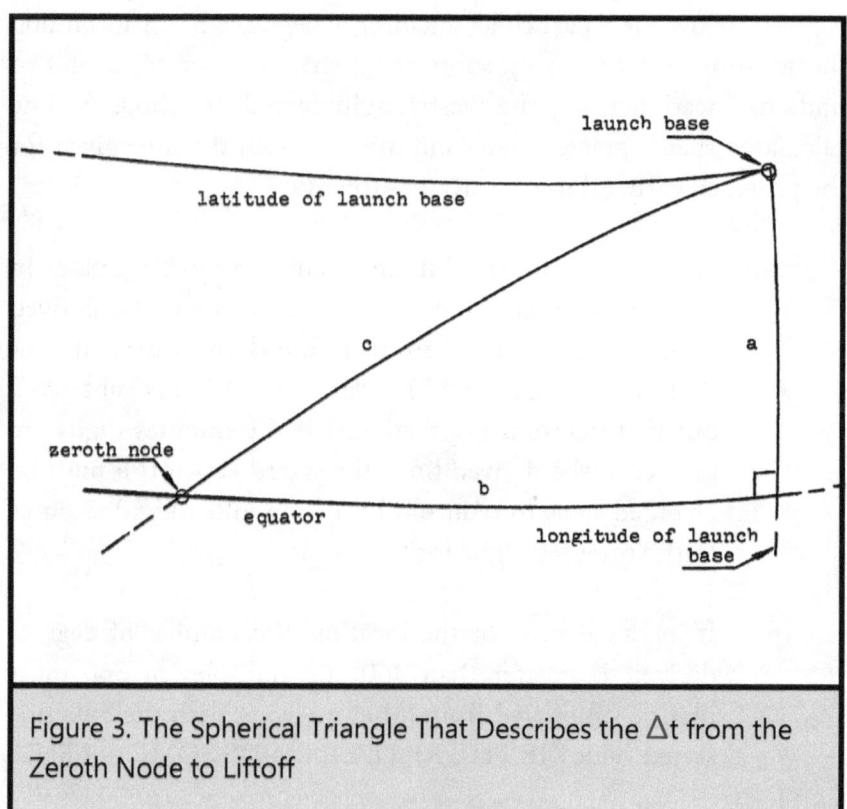

Figure 3. The Spherical Triangle That Describes the Δt from the Zeroth Node to Liftoff

Chapter 3: When Are They Launched

- From the bulletin the zeroth (ascending) node is known.

- The latitude and longitude of the launch base are known.

- Therefore, b and a are known.

- To find c, the spherical triangle formula is used, (for a rt. sph. Δ)

    $$\cos e = \cos a \cos b$$

- After finding c from this in degrees, make the following calculation:

    c/360 x satellite period in minutes = time to traverse the arc which is equal to some time Δ t.

- Add Δ t to the time given in the bulletin for the zeroth node and the result is the liftoff (or launch) time.

An example of the determination of the zeroth node when the bulletin starts with some node other than the zeroth.

A real case can be shown with the data for *Cosmos 393*, launch date Tuesday, January 26, 1971, at an inclination of 71°. From bulletin 2 for this launching we read;

Rev	Time Z	Long W	(note: here rev is identical to orbit and Z=GMT)
4	1840.05	89.64	
5	2012.21	112.91	

Then proceed as follows;

First subtract orbit 4 time from orbit 5 time, thus

```
 2012.21
-1840.05
```

Note that the minutes digits for orbit 4 are greater than those for orbit 5, hence rewrite as,

```
2012.21 = 1972.21
```

then,

```
 1972.21
-1840.05
 -------
 0132.16,
```
giving an orbital period of 1 hour 32.46 minutes or 92.16 minutes

Since we commence with orbit 4 we have (92.16) x (4) 368.64 minutes. Reconvert to hours and minutes in the form presented in the bulletin, thus the time is 0608.64. Subtract this from the time for orbit 4.

```
 1840.05
-0608.64
 -------
 1231.41 z,
```
the time for the zeroth node (on January 26).

For the longitude location:

Subtract the two values given in orbits 4, 5, thus,

112.91° - 89.64° = 23.27°; this is the rotational displacement of the ascending = node per orbit, all factors considered.

Then, (23.27°) x (4) = 93.08°. Subtract the latter from the orbit 4 location,

89.64° -93.08° = -3.44°

Since we are calculating in west longitude, a minus sign means east longitude so that 360° - 3.44° = 356.56° west longitude. From history one will recognize whether or not the west longitude of the zeroth node is in fact correct. We look at Table II for confirmation. The values are close enough to be satisfactory.

I would like to note that the values do not have to come out exactly like those in Table II, a value within 1° is satisfactory and sufficiently indicative that the calculation is a good one. If the west longitude for a given base and inclination does not agree with Table II by some substantial amount then you can be certain that the calculation is wrong. The time will be wrong as well.

If a new inclination is introduced by the Soviet Union then some calculations will have to be made for several launches at this new inclination using the spherical trigonometry previously noted. The values will have to be averaged, time and longitude as well, and these inserted in Table II to be reviewed as a base for future calculations with respect to this new inclination. The method is more than ample for obtaining liftoff times within a few minutes of the true time. Users of the technique can apply it and make comparisons with announced times (by the Soviets) when those opportunities arise.

Table II: Approximate Time Interval, Δt, for the Arc Length from the Zeroth Node to the Location of the Launch Base for Each Base and for Each Inclination[25]

Zeroth Node West Longitude Degreees	Orbit Inclination Degrees	Δt Minutes	Launch Base and Its Geographic Location
314	71.3	7.5	Tyuratam
321	65	8	45°55.3' N
326.6	62.25	8.5	63°20.5 E
336	56	9.2	
347	52	10.4	
353.8	49.6	16.4	
1.74	69.2	18.3	Plesetsk
14.4	65.8	19.5	62°54' N
16.5-18	65.4	20	40°39' E
22	64.6	20.5	
45.4	62.8	21.5	
346-8	74	16.5	
352	72.9	17	
356.7	71	17.5	
326	83	14.5	
333	81	14.6	
38	49	20	Kapustin Yar
47	49	20	48°20.25' N
59.3	50.6	26	45°51' E

---

25 Handwritten note on original document: "to be updated, change page to newer table."

4

TO THE PLANETS

In the early years of the space endeavors of the Soviet Union, many of us assumed that her move into space would be a continually expanding sphere, encompassing ever increasing portions of the solar system and crowned with a goodly degree of success in her interplanetary efforts as had, and continued to be the case, with Earth-orbiting satellites. That euphoria was soon scattered by the winds of disaster, which blew with appalling regularity, at each and every planetary window. The first notice that the general public got that the Soviet Union was not limiting herself to the vicinity of the Earth-Moon system came on Sunday, February 12, 1961—just two months before the impact of the first man in space was to amaze the planet, and shake the complacency of the United States to its very roots.

Venera 1 was launched at about 0209 GMT from Tyuratam and demonstrated what was to be <u>the technique</u> for launching spacecraft into translunar or interplanetary trajectories. The method involved placing a satellite into a "parking orbit" around the Earth from which the spacecraft is launched—time for payload checkout and

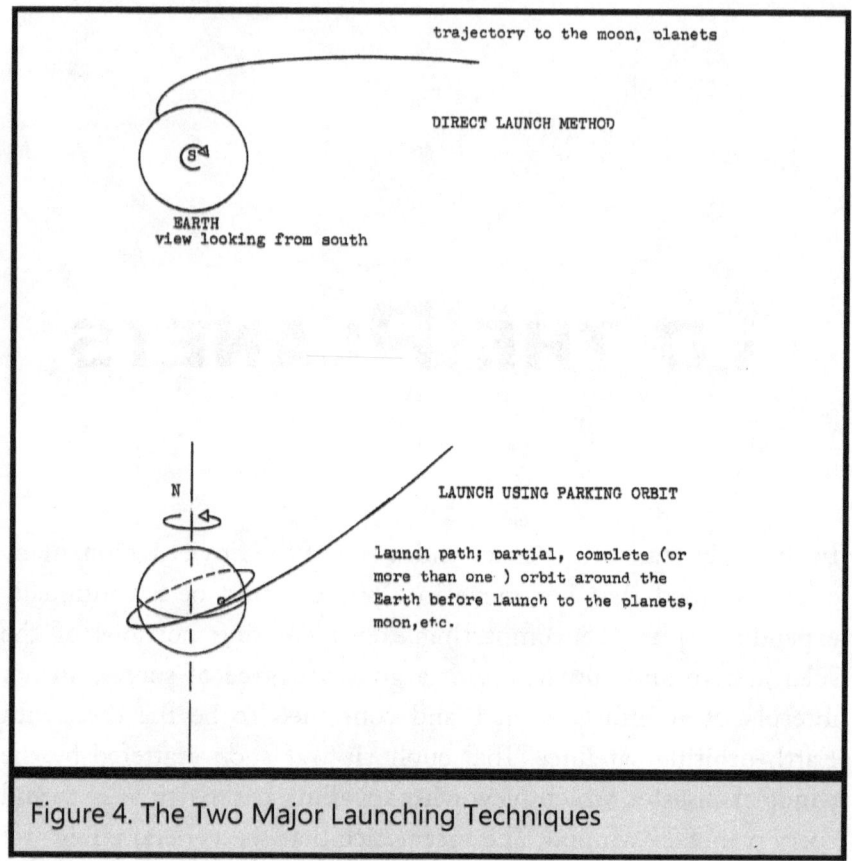

Figure 4. The Two Major Launching Techniques

choices of launch geometry are attained in this manner and most important the criticality of launch time is eased.[26]

In the Soviet's application of this method, the spacecraft circles the Earth for not quite one orbit before planetary injection. The last stage of the booster having completed the task of placing the probe together with its "injection" stage into orbit then separates; telemetry from the spacecraft now informs the ground crew of the status of the bird. Then, at a preset time the injection stage is ignited, burns for a fixed interval, and adds the essential velocity increment to the bird

---

26 The direct method of launching used for *Luna 1, 2,* and *3* as well as the parking orbit method are depicted in Figure 4.

after which it leaves Earth orbit and proceeds on its trajectory to its intended target planet. The ignition usually occurs over equatorial Africa as the spacecraft is some 70-80 minutes into its first (and only) Earth orbit.

There is some variation in the timing because of the variable geometry that may exist among launch windows to the planets and to the Moon. Whereas a satellite in Earth orbit requires some 25,000 feet per second to maintain its orbit, an interplanetary vehicle needs substantially more to send it on its way. In the case of *Venera 1* this was about 38,000 feet per second (fps). There are three classes of velocity as described by the Soviets; the first "cosmic" velocity is that necessary to place a satellite in Earth orbit, the second to escape completely from the influence of the Earth, and the third to escape entirely from the solar system (to escape the influence of the Sun). To just escape from the Earth regardless of direction into the solar system and ignoring the extra velocity to overcome atmospheric drag takes 36,175 fps provided the spacecraft has first been placed in an Earth orbit at an altitude 115 miles above the Earth's surface.[27]

In their announcement the Soviet Union stated that *Venera I* had achieved a velocity of 661 meters per second (m/s) (2168 fps) over the second cosmic velocity so that the total velocity was 38,343 fps assuming that the injection was, indeed, from a 115-mile-high orbit. It could not have been much different; a much higher orbit would be wasteful of energy, a much lower one would yield considerable aerodynamic disturbances even in a temporary parking orbit.

The publicity given this launching—no doubt with a thought of total success overriding—was literally enormous in the Soviet press (as well as in the world press) and practically constituted a short course in interplanetary flight. From their unquestioned position of

---

27 At higher altitudes the velocity required for escape decreases.

superiority in space activity at that time they were evidently certain of success to come. It was not to be. Interplanetary flight was to demand an entirely different level of sophistication, well beyond the needs for short Earth orbit flights, including vastly increased reliability, quality control, precision in manufacturing, and especially in preflight testing. They were not to accomplish this for a long time. Witness, if you will, the same problem in the first six flights of the US Ranger Program. The full realization of this need for increased attention was not to manifest itself for several long years.

I reflected on a launching of just a week earlier, called *Sputnik VII*, which had been announced as a new heavy satellite, but whose purpose was not given. Its weight was announced as 14,293 lb—very impressive. It dawned on me that this had been a first attempt to launch a Venus probe but had not gotten out of Earth orbit. A propulsion failure, or one in guidance and control, or a failure of some crucial instrument aboard, had prevented interplanetary injection. No explanation was ever offered on the mission of *Sputnik VII* but the omission was so obvious that it was relatively simple to conclude the intended use. Because planetary windows are well defined, the time for launching during each day of the window can be determined with some reasonable accuracy without a great deal of calculation.

The *Sputnik VII* launch time of about 0224 GMT on February 4, 1961, fitted well since the launch time for a particular window for launch to Venus grows earlier each day by several minutes for launches by the Soviet Union from Tyuratam. The timing here was about 2 minutes earlier per day between the launches of *Sputnik VII* and *Venera 1*. The times for this pair of launchings is not nearly as accurate as those times obtained later on downstream (a few years later) when data became more readily available and itself more accurate. Additionally, I made some calculations to determine whether or not the weight given for *Sputnik VII* was sufficient to launch a probe with the announced weight of *Venera 1*, 1,418 lb,

Of course it was, even with conservative choices for the propellant energy and structural weights.

I knew the required velocity to good accuracy—given by the Soviets themselves for *Venera 1*—so that all of the needed information was at hand for such a calculation; whatever was not known for sure could be very well bounded. Their second try was not very fruitful either for some 15 days later on February 27, the Soviet Union announced that contact with the probe had been lost. No specific malfunction was announced. Contact with the probe had been maintained on a frequency of 922.8 MHz and some studies of the environment of interplanetary space were conducted. *Venera 1* was presumed to have passed Venus on May 19-21, 1961, at a distance of fewer than 62,000 miles.

The learning process for successful interplanetary flight was to take a long time. It is educational to examine the images depicting *Venera 1* and the photos of later Venus spacecraft. The differences in workmanship, even from the photos, are seen to be real. The improvements attest to the diligence, perseverance, and realization of the completely unforgiving requirements for interplanetary flight. With no man to provide maintenance, repair, modification, reprogramming, or to occasionally kick a balky computer, unmanned spacecraft to the planets leave little leeway, indeed, for error from launch to working arrival.

The next set of planetary attempts by the Soviet Union could only have been a nightmare for them, for out of six attempts only one probe was injected into an interplanetary trajectory. Even worse, the public was told of two earlier Soviet failures by James Webb, then the dynamic head of NASA.

This really unhappy tale began on August 25, 1962, at 0253 GMT when the first of three attempts to launch a probe to Venus took place at Tyuratam. The spacecraft was placed in a low Earth orbit, but when the time came for the ignition of the planetary injection stage the probe exploded and scattered bits and pieces into many orbits, none of which, incidentally, lasted very long in orbit. In a matter of days all the pieces decayed.

On September 1, 1962, again a Saturday, just a week later a second attempt was made at 0224 GMT and the dismaying result differed little from the first case. This spacecraft did transmit during its first orbit on 20.005 MHz but like its predecessor blew up when the planetary propulsion stage was ignited. It too sent many pieces into various orbits, all short-lived. Once more on September 12, at 0141 GMT, the Soviets made a third attempt and it also failed. Some seven or eight pieces were detected in orbit; the third attempt was the last one for that Venus window.

I pondered now whether or not the Soviets would make an attempt during the Mars launch window coming up in late October to mid-November 1962. Their three Venus failures certainly must give them pause to do some rethinking on the nature of the propulsion problem. I needn't have pondered. On October 24, 1962, at 1755 GMT a spacecraft was once again launched from Tyuratam. I can recall at least 24 pieces in orbit, the last one of which decayed on February 26, 1963. The first "visible" Mars probe attempt had failed.

On Thursday morning, November 1, 1962, I was pleased to hear that the Soviet Union had successfully launched a spacecraft to Mars. The launch took place at 1614 GMT. Once again, the newspapers were full of details about the proposed interplanetary experiments but not at all clear on what was to take place when the probe approached Mars. I felt reasonably certain that TV photos would be obtained in addition to various radio-wave probing of the atmosphere and surface. The spacecraft was substantially larger than *Venera 1* (1,418 lb) at 1,970 lb. The workmanship as observed in the photos released represented a considerable improvement. I was pleased for the source of new knowledge in the interplanetary area would be most welcome.

As information was collected by the *Mars 1* it was released with little delay by the Soviet Union. Reasonably certain of impending success, data on this successful bird piled up; the frequencies that it broadcasted on were given as 187.36 MHz, 936.8 MHz, 3747.2 MHz, and 5995.52 MHz. The latter two are of the class used in communications satellites in synchronous Earth orbit. As time progressed reports continued to come in on *Mars 1* but somewhat decreased in number. In the May 17, 1963, issue of *Pravda* a new report revealed that the probe had lost attitude control so that the main receiving antenna was no longer pointing at Earth.

Subsequent efforts to regain control were in vain and all contact with the probe was lost and so admitted by Soviet scientists. Contact appears to have terminated on March 21, 1963, so that the probe had an effective responsive lifetime of about 140 days, a real improvement over *Venera 1*'s 15 days. There were 61 communications sessions with the probe during which over 3,000 commands were transmitted, received, and acted on. Communications were held out to 66,340,000 miles by far the longest distance for such activity with any spacecraft from Earth to that date.

While *Mars 1* was in its first few days of travel and the soft glow of success suffused throughout the scientific community, the Soviets tried again on November 4, 1962, to launch another Mars probe at 1535 GMT that Sunday afternoon. This spacecraft suffered the fate of the earlier failures and at least five pieces were seen on orbit; the last of which decayed in early 1963.

Thus, all this activity in the space of a few months yielded only one semi-success. The multiple attempts had given rise to a great deal of noise in the US Congress and during the sessions on US versus

USSR space progress James Webb revealed that not only had the Soviets tried twice in 1961 for Venus and three times in 1962 for Venus, and three times in later 1962 for Mars, but also that they had also tried twice in October 1960 for Mars. On October 10 and 14, 1960, two Mars probes were launched, he said, neither of which even attained Earth orbit. This startling bit of news prompted me to wonder just how much the Soviet Union was spending per year for their interplanetary programs. I assumed that all six 1962 launches were, roundly speaking, 2,000 lb each and that the earlier 1960 and 1961 launches were of the order of 1,500 lb.

Upon the advice of a colleague, I used $10,000 per lb for the payloads, a not uncommon number for that type of hardware and came up with about 200 million dollars for the two years covering those launches. The number is probably conservative since capital investments were not considered at the launch base plus many other related items. Very crudely, I doubled the number as an all-inclusive amount to cover everything from initial designs to launching of the final product. The later much more sophisticated probes must have cost more; however, it seems safe to say that the Soviets expended at least that much during 1971 in their program to investigate the planets—that is, $200 million per year.

The launch window to Venus came into being once more during early 1964. The geometry for such launches from Earth is correct every 19½ months; we were to see the Soviet Union take advantage of every Venus window except that of late 1973 and all the Mars windows except those of late 1966/early 1967 and that of 1975.

The morning of Friday, March 27, 1964, brought forth the news that another Cosmos spacecraft was placed in orbit by the Soviet Union. There was nothing particularly interesting about the news and I didn't realize that there was more to the story than that. Within a few days and the arrival of Space Detection and Tracking System

Chapter 4: To the Planets

(SPADATS) bulletins this spacecraft proved to be of more than passing interest; had been launched at 0325 GMT into an orbit of 119 by 147 sm, at an angle of 64°48', and with a period of 88.7 minutes. It broadcasted (a beacon) on 19.735 MHz and decayed in less than two days without any further comment.

A bit of calculation indicated that it was just right to be a Venus probe attempt. If this were true there would be further attempts during this launch window. Sure enough, on April 2 at 0249 GMT a spacecraft was launched and soon was injected into an interplanetary trajectory; the Soviet Union announced this spacecraft as *Zond 1* (*Sonde*—or probe) but did not announce any destination. This was foolish for a continual stream of bulletins from the Soviets giving the celestial coordinates of the probe made it plain that it was bound for Venus. A day after launching a midcourse correction was made (at 1818 GMT on April 3) to make a small alteration in the trajectory of the vehicle in order to bring it closer to the path that would ensure its meeting with Venus.

On May 14 another midcourse correction was made; this one was substantial at a velocity change of 164 fps. It now appeared that the flight would reach Venus about July 24 after a flight time of 113 days, it could have been as much as a week earlier or later. On May 18 a bulletin was issued discussing the many communications sessions held with the probe on 922.76 MHz and the fact that all was essentially normal. It was the last I heard of *Zond 1*. Evidently, it too had some sort of failure, I never did find out how close the 2,000 lb (my estimate) spacecraft came to Venus. It remains in its eternal heliocentric orbit to be discovered some day in some future flight by an astronaut/cosmonaut explorer who will bring it home to Earth as a reminder of the quaint vehicles used in very early attempts to explore the planets before man himself set out to investigate them firsthand.

True to form, and ever persevering, the Soviets launched *Zond 2* to Mars just after midday on Monday, November 30, 1964, at 1330 GMT. Some 77 minutes later the probe was injected into an interplanetary trajectory under the control of the "long distance communications and control center" at Yevpatoriya on the Crimean peninsula. This occurred just two days after the United States had launched *Mariner 4* and I found it delightful that both space powers were simultaneously on their respective way to investigate the same planet. The parking orbit parameters were the familiar ones, low orbit, 65°, and so on.

Shortly after launching, the Soviets announced that *Zond 2* was having electrical power problems and that only half the normal power was being generated. That was crucial, indeed, for its mission could not be carried out in that state. Somewhere in the first weeks of flight the problem cleared up, the second solar array had deployed properly. When an announcement was made that *Zond 2* carried six plasma-jet engines for attitude control, the scientific community was really surprised. Such engines are indicative of considerable sophistication.

The plasma engines were successfully tested for 10 days, from December 8 to 18, 1964. At other times, conventional gas-jet engines were used to control vehicle attitude.

Again, bulletins were issued rather regularly and again we waited impatiently for results. In mid-February 1965 news ceased and an ominous silence reigned. I really began to wonder if their program would survive this unbroken string of failures. I reflected on the situation that would exist in the United States if NASA had such a long line of unsuccessful planetary attempts; congressional critics up in arms, yammering for someone's skin, and playing their usual parochial politics. There are few members of Congress who have the foresight to support the space program beyond the time that they will serve in Congress. It also occurred to me that sooner or later the Soviets were going to be successful and then there would be no holding their drive into space to just the "local" planets. A success of sorts soon to come would, nevertheless, prove to be almost annoying considering the circumstances.

On February 25 an announcement from Jodrell Bank in England stated that they had last picked up signals from *Zond 2* on February 17. In a statement from Houston, Dr. Charles Sheldon, who was then a member of the National Aeronautics and Space Council, said that Mstislav Keldysh—then President of the Academy of Sciences of the Soviet Union had just told him that *Zond 2* would pass Mars within 930 miles (the observation should be made here that nothing was said of the status of the spacecraft, so that the close pass might not be as meaningful as first supposed).

Much has been made of the presumed probability that *Zond 2* was intended to land on the Martian surface (or send in a probe to sample the atmosphere). This has been discussed in the magazine, *Science,* in two articles where the authors are concerned with the contamination that would have occurred had *Zond 2* impacted on Mars.[28]

The question remains somewhat unresolved. In discussions with Keldysh both Sir Bernard Lovell and Dr. Sheldon were told that *Zond 2* would fly by Mars during the period of August 6-9, 1965, at a distance of 1500 km (930 sm). Sir Bernard is convinced that *Zond 2* did not impact on the surface of Mars, based on the remarks Keldysh made to him.[29]

---

28 "Planetary Contamination I: The Problem and the Agreements," N. H. Horowitz, R. P. Sharp, and R. W. Davies, and "Planetary Contamination II: Soviet and US Practices and Policies," Bruce C. Murray, Merton E. Davies, and Philip K. Eckman, *Science,* March 24, 1967, vol. 155.

29 "Letters," *Science,* August 4, 1967, vol. 157.

All of this speculation, this intriguing solar guessing game, came to an end on May 5, 1965, when Gennadiy Skuridin, a Soviet delegate to a Chicago conference on space exploration, stated that, "Transmissions from *Zond 2* have stopped, we have not been able to raise it again." So ended the saga of *Zond 2*.

Somewhere in the course of all of this stream of unsuccessful planetary probes, a team of persons was assembled to examine, with care and at length, the reasons for all of the malfunctions insofar as they could be determined. Such an investigation could come up with some very educated guesses as to the source of the troubles using spare spacecraft at their flight centers and manufacturing plants, and by detailed examination of telemetry received from the actual spacecraft in flight. Using that information an experimental probe could then be assembled and instrumentation put aboard it to examine and radio back every detail of the spacecraft's functioning during flight. This is completely analogous to a total physical exam by one's doctor, and, in fact, the spacecraft instrumented this way is called a diagnostic spacecraft.

On Sunday, July 18, 1965, the Soviet Union announced the launching of *Zond 3*. It was launched at 1436 GMT, entered a parking orbit with the by now familiar characteristics; 130 by 101.6 sm, 64.78°, and with a period of 88.42 minutes. Some 75 minutes later it was injected into a planetary trajectory BUT AIMED FOR NO PLANET IN PARTICULAR. No launch window was open for either Mars or Venus and the Soviets made this clear in a statement. The spacecraft was the largest one yet sent into a planetary orbit, it weighed 2,116 lb, close to four times the size of *Mariner 4*, the US Mars spacecraft. Some 33 hours after launching *Zond 3* turned a single camera on a portion of the Moon that had heretofore not been photographed and proceeded to obtain a series of excellent photos of the hidden side along with some photos of the visible side. The details of the portion of the Moon photographed are shown in Figure 5.

On the right of the -90° meridian is located part of the visible side of the Moon covered by the photographs taken by *Zond-3*. On the left, at a longitude -166°, is the morning terminator, the boundary between the lit and unlit parts of the Moon on July 20, 1965. Noted here is the boundary of the region of the obverse side of the Moon photographed by *Luna-3* in 1959. The region photographed by *Luna-3* extends to longitude 110° and is beyond the limits of the sketch herein.

* * *

Photographs commenced at 0124 GMT on July 20 and continued until 0232 GMT the same day. During the procedure altitude above the lunar surface ranged from 7208 sm to 6214 sm. The TV equipment was good having 1100 lines per frame (compare that to home TV with 525 lines) and the pictures were transmitted using a highly directional parabolic antenna. An article in *Pravda* of August 17, 1965, said that the intent was to continually send the photos from the spacecraft to Earth as long as the spacecraft was operating. Because of the desire to perform this procedure from distances exceeding 100 million miles, the video transmission system was rigged to send one photo every 34 minutes. Power limitation, of course, dictated the low rate.

Transmissions were initiated at 1,364,000 miles and continued for months. However, while the probe was close to Earth the 25 pictures taken were transmitted at a rate of one every 2¼ minutes. This rapid scanning of the onboard stored photos resulted in a loss of resolution but were still satisfactory for the experiment being conducted. The photos received were as good as photos taken from Earth via telescope of the visible side of the Moon. A spectrum of experiments, by now a standard practice, were conducted in studies of magnetic properties of near-Earth space and the interplanetary medium, of the solar wind, low frequency radio waves from our galaxy, micrometeorites in space, cosmic rays, along with infrared (IR) and ultraviolet studies of the lunar surface. However, the eyebrow-raising experiment was the use, again, of plasma-jet engines for attitude control and this time for making course corrections for the probe. A midcourse correction was performed on September 16; the correction was purely experimental and equaled 164 fps. No information was released on whether or not the plasma-jet engines participated in this correction. For the purpose of orientation, the Sun and the bright star Canopus was used. Canopus seems to be everybody's favorite star—it is used frequently in US space probe missions.

By late October 1965, 99 communications had been held with the probe and it was now 20 million miles away. The last Soviet bulletin on *Zond 3* was published in *Pravda* on March 3, 1966; it stated that 135 communications sessions had been held with *Zond 3*, that photos of the Moon were still being received and, "a large volume of scientific information as well as data on the operation of the onboard systems were transmitted to Earth." The probe was at a distance of over 95 million miles on that date and had operated faultlessly for 227 days.

Though pleased with the performance, the Soviets must have suffered some chagrin, for here an experimental probe had operated for far more than that necessary to reach Mars and for more than twice the time necessary to reach Venus.

Before the experience with *Zond 3* was over, a new launch window for Venus rolled around on the calendar during November-December 1965. I felt confident that new launchings would occur. And so they did. On Friday morning at 0447 GMT on November 12 the spacecraft *Venera 2* was launched into a low Earth orbit and something new was added. The orbit was a bit higher than usual at 162 by 144 sm with a period of 89.11 minutes but with a new inclination of 51.87°. The lower inclination did two things for the Soviet scientists of the space program; it gave the probe an added increment of the rotationally derived velocity due to the spin of the

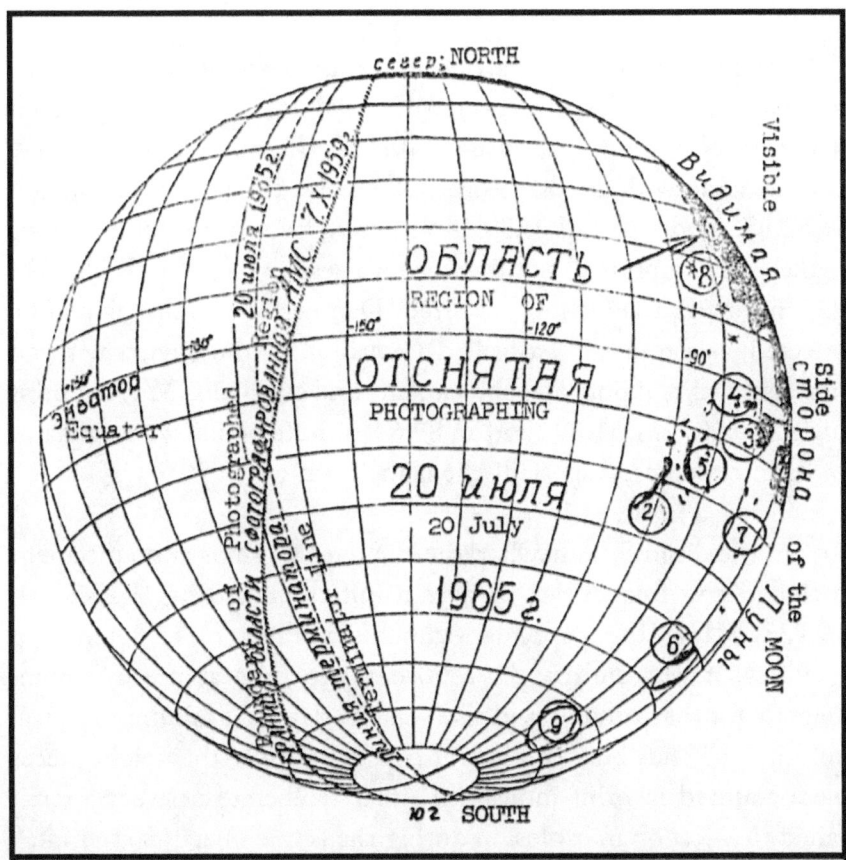

Figure 5: Diagram of the location of the domain photographed by *Zond-3*

1 Ocean of Storms
2 Eastern Sea
3 Crater of Grimaldi
4 Crater of Riccioli
5 Sea of Autumn and Sea of Spring
6 Schickard Crater
7 Crater of A. Biruni
8 Crater of A. Einstein
9 Crater of Bayer.

Earth and, because the maximum latitude excursion of the probe while in Earth orbit was considerably less than that exhibited in the 65° orbit, it allowed the controllers at Yevpatoria to keep the probe in radio sight for a much longer period. Both of these are advantages. Moreover, no disadvantages accrued from this change.

All interplanetary spacecraft were to be launched at this new inclination from the initiation with *Venera 2*. Four days later on Tuesday, November 16, at 0413 GMT, *Venera 3* was launched, commencing with an Earth orbit of 188 by 131 sm, an inclination of 51.87°, and a period of 89.64 minutes. Two successful injections looked very impressive but then came Tuesday morning, November 23, and a third launching occurred. Damned if it didn't fail to be injected; as a result, it was called *Cosmos 96* by the Soviets with no hint as to its real purpose. The launch time of 0313 GMT, the orbit of 182 by 137 sm, 51.89°, and an 89.64-minute period were evidence of its intended use, especially the launch time.

To aid the disguise somewhat, the standard Cosmos announcement included the frequencies of a transmitter aboard at 19.895 and 19.735 MHz. The particular launch date had other interesting qualities; it was the day for a 90-day trip time at the minimum velocity for that window and also was the day for a minimum arrival velocity at Venus but for a longer trip time. More than eight pieces were counted in orbit indicating either deliberate destruct after it failed to inject or an explosion during the actual injection attempt.

Venera 2 weighed 2,123 lb and *Venera 3* weighed 2,116 lb. The former was intended to fly by the planet at a distance of about 15,000 sm. No midcourse corrections were necessary for *Venera 2* for its flyby, which indicates the accuracy at the original injection; recall that at burnout the velocity is close to 38,000 fps. As the Soviets state, a 1 meter per second error in velocity (3 fps) yields an error at the planet of over 18,000 miles in position at flyby. The *Venera 3* spacecraft required a midcourse correction, for its intended target was Venus itself. On December 26, 1965, a midcourse correction was imparted to *Venera 3*.

At regular intervals bulletins were issued on the progress of both of the probes. Finally, after 3½ months in space the following announcement was made by the Soviet Union through its news agency TASS in the newspaper *Pravda* of March 2, 1966.

> At 0956 Moscow time (0656 GMT) of March 1, 1966, the Venera 3 probe reached the planet Venus after a flight of 3½ months in space, and placed a flag with the seal of the USSR on its surface. The exact encounter of the probe with the planet was guaranteed by a successfully executed correction of the probe's flight trajectory on 26 December 1965. During its flight regular radio communication with the probe was maintained and scientific information was received. There was no communications session in the concluding stage of the probe's approach to the planet Venus.
>
> The other probe, *Venera 2*, launched on November 12, 1965, continues to fly in a heliocentric orbit after having passed within 24,000 km (14,880 sm) of the surface of Venus at 0552 Moscow time (0252 GMT) on 27 February 1966. The probe's flight at a given distance from the planet without any intermediate correction was guaranteed just by its exact injection onto the interplanetary trajectory.
>
> The experiments performed with the aid of the probes Venera 2 and Venera 3 permitted the solution of a number of new, in principle, problems of interplanetary flight, and also the obtaining of new scientific data. The flight materials of these probes are being processed and studied.

For the record, the *Venera 2* flight took 107 days, 22 hours 6 minutes to its close pass at Venus, while *Venera 3* took 105 days, 2 hours 43 minutes to its landing on Venus.

The emphasis on the lack of communications is mine. Evidently, the great heat of the planet together with the increased heat received from the Sun at that much closer distance than Earth, were too much for the *Venera 2* probe. Those circumstances plus the high-pressure furnace that is Venus itself proved to be too much for *Venera 3*. Some information indicating the increasing temperatures was undoubtedly telemetered back to the Soviet Union. Just "watching" the engineering data of the spacecraft as they approached the planet would have been sufficient to obtain such information. Future events were themselves evidence that *Venera 2* and *3* provided much learning toward exploring the planets. Next time things would go better.

The day after the unhappy notice in *Pravda* that the Venus probes were not successful, the TASS report on the great success of *Zond 3* was published. There was a contrast in results that make scientists and engineers grow old before their time.

History would soon soften the blow. Perseverance was about to pay off. None of us knew, how could we, that the next Venus launch window would, indeed, be different.

The summer of 1967 rolled around and we rubbed our hands in anticipation; both space powers were going to launch to Venus, new knowledge was bound to be obtained. And unknown then, a roaring international scientific argument was in the making.

Venera 4

The morning of June 12, 1967, was a Monday, a day of the week for which I have no great love. But this one was brightened considerably with the news that *Venera 4* had been launched at 0240 GMT and was safely on its way to Venus. The Russians must have felt fairly good about this launch also, for their first announcement stated the launch time and the probe's weight of 2,436 lb (1,106 kg). The launch was from Tyuratam at the new inclination, 51.778°, started with *Venera 2, 3*. The parking orbit was 132 by 106 sm. The probe weighed more than 300 lb greater than each of the two previous birds.

The Soviets tried to launch again at 0235 GMT on June 17 but the propulsion bugaboo plagued them once more and the failed bird in Earth orbit was called *Cosmos 167*.

Despite the optimism that seemed to reign, the Soviets were cautious in their running comments on progress of *Venera 4*. On July 15 it was stated that by July 14, twenty communications sessions had been held with the probe and the probe was now 5.1 million miles from Earth. On July 29 a midcourse correction was made with *Venera 4* when it was some 7.4 million miles from Earth; however, this piece of information was not released until September 16 by which time there was great assurance that the probe was on course. I announced to the press that this would be a soft landing on Venus (*San Jose Mercury*, Tuesday, October 3, 1967); so sure was I of success of the probe. The *Pravda*, September 16 article, also noted that the probe was now 23,850,000 miles from Earth and that 91 communications sessions were held with the probe. There was one more long month to go before that glorious meeting with that planet-sized hell called Venus.

Now it was October 16, 1967, and a real, 24-carat bombshell hit the scientific world: Mstislav Keldysh, President of the Academy of Sciences of the Soviet Union, sent a telegram to Sir Bernard Lovell asking for aid in tracking *Venera 4* and taping its data as the spacecraft neared its target planet. Sir Bernard was delighted and hailed the request as "the first real sign of a breakthrough in international cooperation with the Russians."[30] In his message, Keldysh stated that *Venera 4* would reach Venus within two days. He stated further, "information about the physical properties of Venus was of primary interest for world science and goes beyond the importance of one nation's scientific experiment." Continuing, the message went on that the experiment was of "extraordinary importance and significance for mankind."

US radio telescopes were probably not in the proper location for receiving the data at the appropriate time although US equipment in South Africa and Australia might have been. The Soviet Union was not about to lose priceless information from *Venera 4* in the unlikely event that its own receiving equipment failed at the crucial moment. The Soviet Union also requested aid from the French but there was considerable doubt that France's equipment would be sufficient.

As for Sir Bernard, his staff and his 250 ft diameter telescope, he made the cryptic remark, "We will be ready." And so he was! *Venera 4* entered the atmosphere of Venus at 0434 GMT on October 18 after a 129-day flight. The entire *Venera 4* spacecraft entered the atmosphere of Venus. There was no need to be concerned about sterilization for that level of intense heat is detrimental to any life as we know it on Earth. Before entry the bus (the whole vehicle whose

---

30 *New York Times,* October 17, 1967.

mission it was to deliver the entry spacecraft) separated the 1-meter diameter entry capsule, which was very substantial, indeed, at 844 lb. This was 304 lb more than the entire US *Mariner 5* spacecraft, which flew by Venus the next day (October 19).

At entry into the Venus atmosphere, the heat generated in the shock wave ahead of the spacecraft was 18,000-20,000° degrees Fahrenheit (F); the entry velocity was 33,440 fps, and the entry acceleration was over 300 g (i.e., over 300 times the acceleration due to gravity at the Earth's surface). There were two braking chutes activated after the aerodynamic acceleration period was over. The latter is a matter of minutes and the 300 g peak probably is over in a few seconds at most. A drogue chute, out first, served to stabilize and brake the descending capsule, which was still moving at almost 1000 fps before drogue initiation. The drogue having slowed the capsule farther then pulled out the main chute and the capsule then moved deeper into the Venus atmosphere at a few meters per second.

The results obtained by the *Venera 4* spacecraft showed that the atmosphere was 90 percent carbon dioxide within an accuracy of 10 percent that about 1 percent of oxygen was contained in the atmosphere, that nitrogen may be some 7 percent of the atmosphere, and that water was very scarce at some small fraction of 1 percent. All of this data was sent over a period of 94 minutes and 50 seconds, after which the spacecraft presumably reached the surface of Venus.[31] But many incompatibilities existed when the *Venera 4* data was compared with the data obtained from *Mariner 5*. The crux of the matter was whether or not the Soviet spacecraft had actually reached very close to the surface while broadcasting data for the temperatures, pressures, and radius of Venus as the data was in disagreement with the American data. Many technical notes were published, followed by lengthy papers in the best and most prestigious journals, and

---

31   Signals were received at the Soviet long-distance communications center from 0439:10 sec GMT to 0614 GMT (October 18).

both of these were intermixed with some strong verbal commentary throughout the scientific community. A real brouhaha!

I got into the act by aiding a colleague who was puzzled about some values published in *Pravda* and *Izvestia* (they gave different values for the same event in the *Venera 4* descent data). I wrote to Keldysh who stated in a very prompt answer that the newspaper data consisted of first results and that the true value lay between the two. The controversy finally focused on one point; at what altitude had data first been collected? The Soviets stated that their altimeter had said 26 kilometers (some 16 miles); others claimed that the data fit an altitude of 52 kilometers and that the Soviet altimeter readings had an ambiguity that they missed in their evaluation of the data.

At first the Soviets were adamant about their readings, then all at once reference to the altitude vanished from their remarks, the ambiguity had, indeed, been found—52 kilometers was correct. The correct interpretation of the *Venera 4* data insofar as altitude and pressure were concerned was that the spacecraft was crushed by the huge pressures at an altitude of about 14 miles, where the pressure was about 22 times the Earth's sea-level pressure—about 323 lb per square inch—and the temperature was 536 degrees F. Data collection had been initiated at an altitude of 32 miles where the pressure was about half an (Earth) atmosphere and the temperature was about 100 degrees F.

In sum, one can see that Venus is a mighty inhospitable place and it will take the most sophisticated equipment to land there and survive long enough in order to obtain video photos and complete some sampling of the surface. To say that the photos will be exotic is a vast understatement. But unmanned, it will be done in not too many years.[32] As for man's landing there, we will have some years to wait.

---

[32] Success in this area came early. In October 1975, *Venera 9* and *10* returned excellent panoramas from the surface of Venus via video.

Chapter 4: To the Planets

Of course, it may never occur but that would be by choice for the advance of technology will inevitably lead us to the point where we will be able to do so if we choose to.

A few days later on October 30, the Soviets brightened my birthday by rendezvousing and docking two unmanned spacecraft, *Cosmos 186* and *Cosmos 188*. These were tests for operations in the Soyuz manned spaceflight program, which was to rack up a few firsts within the following years, but more about that in Chapter 8, "Soyuz."

Early in the morning (GMT, that is) when the temperature was a frosty 30 degrees F below zero as reported by the Soviet Union on Sunday, January 5, 1969, at 0628 *Venera 5* was launched from Tyuratam. I examined the velocity—contour maps prepared for NASA by Dr. Stanley Ross and observed that the Soviet space program people had selected a good day, indeed, for the launch for the maps showed that this was a time requiring a minimum velocity for the launch window. Burnout velocity was 37,433 fps and the trip time looked like it would be close to 130 days.

Once again, the spacecraft weight was increased; this time by 55 lb. The probe weighed 2,491 lb. The parking orbit at Earth was the by now familiar type; 122 by 114.4 sm, at 51.786°, and with a period of 88.28 minutes. As usual the spacecraft was injected onto its interplanetary trajectory before one orbit was completed leaving various shrouds and the last (third) stage of the booster vehicle in Earth orbit. Feeling more confident now, the Soviets announced that the injection had occurred 79 minutes after liftoff, at 0747 GMT. They went so far as to state that the burning time of the injection stage was 228 seconds.

Since the injection velocity was largely known from the contour maps, the payload was given, and the propellants could be reasonably guessed, I made some calculations on the weight of the empty rocket stage that went along toward Venus with *Venera 5*. It appears to have been about 2,500 lb. I estimated the specific impulse to be about 320 lb/sec so that the engine was found to burn about 40 lb of propellant per second and therefore the thrust of the engine was calculated to be 12,800 lb and some 8,940 lb of propellant was used to attain the injection velocity of 11,850 fps.[33] All of these values are educated guesses except for those released by the Soviet Union so that if these answers are within say 10 percent I would be pleased. Some day we may even be given the exact values by the Soviet Union but we'll have to wait until their confidence in themselves increases enough for the unnecessary veil of secrecy to be dropped.

On Friday morning, January 10, *Venera 6* was launched from Tyuratam at 0552 GMT under conditions similar to those for *Venera 5*. I estimated a flight time of 127 days using the contour maps (this was to prove to be exactly right whereas I missed the *Venera 5* arrival by four days). Both spacecraft were broadcasting on 922.763 MHz and I assumed that they were far enough apart on their flight paths so that signal separation could be maintained or that they had different modulations in their signals so that they could be distinguished.

Both spacecraft were to enter the atmosphere of Venus and make a soft landing on its surface, collecting temperature, pressure, and atmospheric composition measurements during (and after) the descent. The information release from the Soviets was improving for here they parted their curtain a bit to clearly state the mission intent right from its beginning. How nice! By the middle of March both probes had midcourse corrections completed, each probe had had more than 25 communications sessions and all was going well.

In orienting the spacecraft for the midcourse correction the Soviet scientists had used Sirius (differing from the US use of Canopus)

---

33 See Appendix C for the equations used in the calculation.

and had used the Sun itself as "aiming" objects. Each for his favorite star. The correction for *Venera 5* was 30.18 fps (9.2 m/s) and was made on March 14 while the correction for *Venera 6* was 122.67 fps (37.4 m/s). The accuracy of the corrections can only be described as fantastic, being one-tenth of 1 percent for each of the spacecraft (1 cm/sec for *Venera 5* and 3 cm/sec for *Venera 6*). The uncorrected injection-rocket stages went into heliocentric orbits.

One hundred and thirty-one days after launch *Venera 5* entered the atmosphere of Venus at 0601 GMT on Friday, May 16, 1969. The next day *Venera 6* entered the Venus atmosphere at 0605 GMT, just 127 days after its liftoff. Both of the entry probes were heavier than their predecessor's—the *Venera 4* probe. They each weighed in at 992 lb versus the 844 lb of the earlier explorer. Because they arrived at a higher velocity than did *Venera 4* the accelerations they experienced were considerably greater at 450 g. However, in an attempt to get them to the surface faster than the descent time of *Venera 4* the parachute size had been reduced "several-fold," thus a 51-minute communications session was held with *Venera 5* and a 53-minute session with *Venera 6*.

Although the internal temperatures of the probes changed from just 55 to 82 degrees F, the external temperature went up to 576 degrees F. When the pressure was 0.6 atmosphere and the temperature 77 degrees F the first measurements were taken by *Venera 5*. At 5 atmospheres and 302 degrees F the second set of measurements were taken. For *Venera 6* the first sampling was at 1 atmosphere and 140 degrees F and the second at 10 atmospheres and 437 degrees. Each probe made more than 50 temperature and more than 70 pressure measurements during the descent. While *Venera 4* had made measurements from 0.5 atmosphere and 77 degrees F down to 18 atmospheres and 500 degrees F, the *Venera 5* and 6 probes measured pressures from 0.5 to 27 atmospheres and temperatures between 77 and 608 degrees F. Together *Venera 5* and 6 found the composition of the atmosphere to be carbon dioxide 93-97 percent, oxygen less than 0.4 percent, nitrogen plus other (undefined) inert gases 2-5 percent, and the water content was 4-11 milligrams per liter (roughly, one to two five-thousandths of an ounce per quart of atmosphere).

Now a peculiar event was noted: both probes sent their measurements while descending through some 37 kilometers (23 sm) of the atmosphere yet their final altitude readings both taken at the same temperature and pressure readings gave quite different results. *Venera 5* read out that it was at an altitude of 15.5-16.1 sm while *Venera 6* read out an altitude of 6.2-7.4 sm above the surface. Since the probes had entered the atmosphere two hundred miles apart it is possible that one descended over a mountain range thus reading its altitude above the "surface" from the mountain height while the other descended over a valley and gave its readings from that lower surface.

The facts are not fully known in the sense that we have no picture of the topography of Venus. Accepting the values as factual would lead to the conclusions that at the surface measured by *Venera 6* the temperature (on the "mountain") would be about 750 degrees F and the pressure 60 atmospheres while the "valley" measured by *Venera 5* would be 985 degrees F and the pressure 140 atmospheres. In both cases, man would be fried like an egg and crushed into a pancake. Very unsatisfactory prospects. Thus, pointing up, once again, the difficulties man will have getting down to the surface of Venus for firsthand exploration.

Like the tides and the trains, launch windows wait for no man. Count 19½ months after one Venus launch window and the next one comes up. So did one during August 1970 and along like clockwork came *Venera 7*. At 0538 GMT on the morning of Monday, August 17, 1970, the spacecraft was placed in a parking orbit of 125.7 by 113 sm, at an angle of 51.77°, and with a period of 88.23 minutes.

At liftoff plus 81 minutes the injection stage was ignited and 244 seconds later the stage shut down as programmed. Again, the payload size increased, *Venera 7* weighed 2,601 lb. It broadcasted on a frequency of 928.429 MHz. Early reports boded well for the spacecraft with all systems behaving normally.

Five days later, on August 22, a second attempt to explore Venus in this launch window was made. The spacecraft failed, however, and therein lies a tale. Liftoff was at 0505 GMT, a correct time by my calculation (for a *Venera* launch). The orbit was announced by TASS as 565.5 by 130.5 sm, with a period of 95.5 minutes, and at an inclination of 51.5°. Well, that was not commensurate with a parking orbit for a planetary probe but the launch time was an odd one for other uses I could think of and it did fit the time for a Venus launching. When I got a look at the *NASA Prediction Bulletin* for the ephemerides[34] for the payload and the last stage of its booster, now both in Earth orbit, the picture cleared up. The last stage of the booster was in an orbit of 191 by 130 sm, with a period of 89.65 minutes, and at an angle of 51.82°. The payload was in an orbit similar to that given by TASS but the angle was 51.23°.

The Soviets were playing footsie; I could well understand their chagrin in this case. Though the angle differences might seem small, as indeed they were, the difference was significant. To begin with, in all of our planetary launches it is necessary to get the payload closer to the plane of the ecliptic than the launching inclination position since the planets of interest lie close to the ecliptic plane—within a very few degrees. It was evident here that this at least started to occur in the launch. I made the appropriate calculations to determine the energy difference represented by the position of the rocket stage and the payload using average altitudes (apogee plus perigee divided by two) and found that this was satisfied by an additional velocity of

---

34 Predicted path for the Earth orbits.

583 fps. Then assuming that the total angle change was 0.6° I found a requirement for another 240 fps. But these two changes occur together so that one has to root-sum-square these for the correct answer (square each value, add them together, take the square root of the sum); the final value was 630 fps.

To get to Venus that August 22 one needed almost exactly an injection velocity of 11,900 fps from Earth orbit and from information on *Venera 7* we know that this took 244 seconds of injection stage burning to obtain. Thus, the would-be *Venera 8* injection stage ignited, burned for some 13 seconds, and then experienced a failure in its propulsion system shutting it down.[35] Damn! I was pleased with my "detective" work but not pleased with the event. I'd much rather a successful launch had occurred; it means knowledge for all of us. There were reports on September 5 that all was well with *Venera 7*. Good. On November 19 a report was issued stating that 85 communications sessions had been held with the spacecraft and that at 0930 GMT on November 17 a midcourse correction was made assuring a meeting with Venus on December 15—the probe was 19.5 million miles from Earth. While *Lunokhod 1* was on the Moon and *Venera 7* was 20 million miles away, both spacecraft made simultaneous measurements of the solar flare of November 19. This sort of concurrent measurement is very useful in understanding solar processes.

Three days before reaching Venus the temperature inside the entry probe was lowered to 18° F and its batteries were fully charged via the solar arrays on the spacecraft. On December 15 the spacecraft reached Venus and entered its atmosphere, roughly at 0502 GMT according to *Pravda* of December 16. "Signals were received for 35 minutes," much less than any of the prior operating probes, and, "information was being processed." The information given out

---

35 See Appendix D for the calculation.

by the Soviets was so brief that many American correspondents decided that *Venera 7* like its earlier fellows had been crushed by the enormous pressures of the atmosphere.

Every planetary window for any selected planet is not exactly like its immediate predecessor and for any launch window there are particular times to launch so that the spacecraft will arrive with "the lowest arrival velocity" for the window in question. *Venera 4* arrived at about 35,500 fps, *Venera 5* and *6* at about 36,500 fps, and *Venera 7* arrived at about 37,800 fps. The acceleration reached in the aerodynamic braking as the probe slammed into the atmosphere also increased each time. These thoughts bubbled around in my head as I wondered what had occurred to interfere with the collection of information by *Venera 7*. The silence by the Soviets was disturbing. Had something unusual happened that was favorable; if so, why wasn't some announcement made? It did and they did.

On January 26, 1971, some 42 days after the event the Russians announced that *Venera 7* had touched down on the surface of Venus and had broadcast from there for 23 minutes. The first thought that surfaced was, of course, how was it known that the spacecraft reached the surface intact. Well, the 42-day interim was spent in using computer techniques to extract the weak (taped) signal from the probe as it sat on the surface. When the spacecraft was constructed very stable broadcasting equipment (frequency generators) were used and were calibrated in flight several times and compared with controls in the Soviet Union. This allowed great accuracy to be obtained in calculation of the rate of fall through the Venusian atmosphere. When a body emitting sound or light or radio waves moves past, a person detects in the case of sound an apparent change in the frequency (pitch) of the sound. With instruments receiving radio waves, the rate of fall through the atmosphere of Venus could be detected.

When the probe touched down that portion of the probe's transmitter's Doppler frequency went to zero. The only Doppler that was then present was due to the relative motions of the Earth and Venus in their heliocentric orbits. This situation was found to exist for the 23 minutes in question. Moreover, when the probe touched down it apparently rolled slightly so that the antenna was not pointed precisely at Earth so that the signal became weaker by a factor of 100 under that received while it was still descending through the Venusian atmosphere. Because the commutator, which samples all of the instruments in a programmed manner in order to transmit data from each of them, became stuck in one position, only the temperature was sampled at the surface. Calculations made using the descent data plus the surface data led to the result that the temperature at the surface was found to be 887 +/- 68 degrees F and the pressure 90 +/- 15 atmospheres. Consequently, the high end of the temperature could be close to 1000 degrees F. Lead, cadmium, tin, and zinc would all exist in the molten state on the surface of Venus.

In the cosmology of the universe, the celestial geometry of the solar system led to the point where the Earth and Venus were again in favorable position for another attempt at the exploration of Venus. This was evidenced at 0415 GMT on Monday, March 27, 1972, when the Soviet Union launched the spacecraft *Venera 8* into its brief Earth parking orbit of 152.6 by 120.3 sm, at 51.79°, and with a period of 88.14 minutes. Eighty-seven minutes later, after a 243 second burn of the planetary injection stage, at 0542 GMT the spacecraft was en route to a rendezvous with Venus.

In an unusual departure for them the Director of the Soviet Institute of Space Research, Dr. Georgi I. Petrov, stated that the mission for *Venera 8* was to soft-land on and analyze the surface and atmosphere of Venus. The 2,610 lb (1,184 kg) spacecraft, like its immediate predecessor *Venera 7*, was to return data on dust, water, and gaseous content of the Venusian clouds. Further, this new craft had been strengthened to withstand both the great pressure and high temperature of the Venusian atmosphere and surface.

After a 117-day trip, during which it returned data on a frequency of 928.4 MHz, entry took place at 38,000 fps, the highest such velocity to that date in the Venera program. *Venera 8* landed on the surface of the planet on July 22, 1972, at 0929 GMT. It then broadcast data for 50 minutes. Prior to landing it broadcasted data during descent through the atmosphere for 56 minutes. *Venera 8* had more of its weight devoted to instrumentation and an increased structural strength of the entry module than did *Venera 7*. Since the overall weight of both spacecraft was the same this series of improvements implied the use of advanced materials for *Venera 8*.

For the first time a *Venera* spacecraft landed on the daylight side of the planet. Since only a thin sliver of Venus, 310 sm (500 km) wide, was both lighted by the Sun and visible to Earth, the en route navigation requirements were quite severe. The trajectory accuracy may be noted by virtue of the fact that only one midcourse maneuver was made early in the flight in April, long before planet-fall.

In order to avoid the problem of *Venera 7*, where the spacecraft did not remain aligned along the local vertical while resting on the surface, and its radio signal attenuated by a factor of 100, *Venera 8* had two antennas, one of which was ejected from its stored position so as to sit upright on the surface to prevent the loss of signal, or its weakening as had occurred with *Venera 7*. Both antennas were used (separately) to return data; first from the main antenna for about ¼ hour, then from the separated antenna for 20 minutes, and then back to the main antenna for the remainder of the broadcast time.

The landing occurred at early morning on Venus. The 1,089 lb (495 kg) entry module separated from the interplanetary bus 53 minutes before entry, which took place at 0833 GMT. The descent took 56 minutes. The initial acceleration pulse at entry reached a maximum of 335 g. In the upper atmosphere between 29 and 21 sm (46 and 33 km), ammonia, NH_3, was found in amounts ranging from 0.01 percent to 0.1 percent. Atmospheric constituents closer to the surface were found to be 97 percent CO_2, about 2 percent N_2, and less than 0.1 percent O_2.

At initial touchdown *Venera 8* had an internal temperature of 68 degrees F. The surface temperature was determined to be 878 degrees F and the pressure 87 atmospheres. Using a gamma-ray spectrometer the surface was found to consist of granitic rocks not unlike those of Earth. While winds aloft at an altitude of 28 miles were found to be in excess of 112 mph, at a 6-mile altitude the winds were low at 45 mph. Soil density was determined to be low at 1.5 gm/cm^3 (recall water is 1 gm/cm^3) and Petrov stated that it was soft enough for the spacecraft to have eventually sunk into the molten surface. Where, no doubt, it was cooked to extinction!

Launch windows occur far too infrequently, much less so than all of us would prefer, hence the great desire to obtain substantial information—new knowledge—at every opportunity within reasonable economic bounds for that part of space program activity. In their efforts to explore the planets the Soviet program, therefore, includes two or more launches at each particular launch window. Such a choice permits data collection from different parts of the target planet thus offering complementary data and a double (or better) opportunity for serendipitous discovery.

The March 1972 launch window reflected that philosophy once again. On Friday, March 31, 1972, at 0402 GMT a would-be *Venera 9* lifted off from the Tyuratam complex. At about 0529 GMT the planetary injection stage was ignited for its intended 243 second burn. However, a malfunction occurred and the spacecraft was transferred from its initial orbit of 133 by 122 sm, at 51.76°, and a

88.54 minute period to one with parameters of 6098 by 131 sm, at 52.2°, and with a period of 201.4 minutes. Using the same method shown in Appendix D (where an incomplete burn for *Cosmos 359* was calculated) the cutoff time for the injection stage was shown to be 93.4 seconds, only 38.4 percent of its intended burning time.

The delta V added as a result of the partial burn was nevertheless substantial; of a needed 12,773 fps, 4,911 fps was added yielding the TASS announced orbit parameters. There has been no clue to describe the reason for the early cutoff. The spacecraft was never identified as a failed Venus probe attempt; rather the Soviets named it *Cosmos 482* and the orbit parameters released were those that existed after the prematurely ended burn. While its perigee has stayed almost constant for over 3 years the apogee of the *Cosmos 482* orbit has been lowered through drag effects to approximately 4500 miles by the end of 1975. The spacecraft will remain in orbit for many years unless, as is fairly likely, some future space shuttle retrieves it for use as a museum piece.

The Venus launch window of late October, early November 1973 approached, came into being, and then receded into history without any effort by the Soviet Union to take advantage of that opportunity. Since the initiation of the exploration of Venus in early February 1961 this was the first gap to occur. There were no faulty launchings. At irregular intervals the Soviet technical literature as well as the daily press had contained discussions and comments on the desirability and the value of obtaining photographs of the surface of Venus.[36] The *Venera 4* to *8* class of spacecraft simply did not have the capacity to collect and transmit high quality video pictures of the Venusian surface; nor in fact were they ever designed to do so. The hostile environment demanded very sophisticated equipment to perform such a task. Further, signal attenuation through the

---

36 Note the predictions made in Chapter 13, "The Future," before the event occurred.

hot dense atmosphere would be best met by broadcasting on a frequency selected for that purpose to a satellite orbiting the planet while broadcast back to Earth was optimized by using a different frequency for that function.

Moreover, relatively simple experiments having been performed with the early *Veneras*, the time had come to conduct more complex investigations and these inherently required increased instrument sophistication. The summation of these new investigatory roles effectively laid down the basis for a new, more complex and, therefore, heavier spacecraft. The omission of the 1973 launch window was evidence that the new spacecraft had not yet gone through its fabrication and testing cycles. I was that convinced that it existed. And it did, for when the following window arrived in June 1975 two new spacecraft, *Venera 9* and *10*, made their now historic appearance.

The Tyuratam complex was thoroughly engaged in the early hours of June 8, 1975, as *Venera 9* was being prepared for launching. The event took place at 0237 GMT and then once more a *Venera* spacecraft was placed briefly in a parking orbit; 123.6 by 107 sm, inclined 51.52°, and with an 88.14 minute period. Injection occurred 77 minutes later at 0354 GMT. Discussion of the spacecraft design details or even its weight was significantly absent from the initial TASS communique on the launching when published in the Soviet press the following day; except for a note that "*Venera 9* is a new type of spacecraft." Such a statement could mean anything from minor changes from prior Veneras to an entirely new craft.

The paucity of detail was itself a firm clue that a really new spacecraft was heading for Venus. Speculation abounded that the Zond booster had been used and this implied a really great increase in weight. Both of these thoughts eventually proved to be correct. When announced in *Pravda* on February 21, 1976, the *Venera 9* weight was given as 10,882 lb (4,946 kg), even larger than the *Mars 2* and *3* payloads. Six days later *Venera 10* also paused for a short time in its Earth orbit at 127 by 102.5 sm, 51.54°, and a 88.13 minute period after being launched at 0259.7 GMT. At 0430 GMT *Venera 10* went through the planetary injection process. Details were again sketchy as with *Venera 9*, leading once more to rampant speculation on its design, specific mission, and size. The *Venera 10* weight was eventually announced as 11,096 lb (5,033 kg), 214 lb heavier than its predecessor. Both were more than four times heavier than *Venera 8*.

En route, *Venera 9* performed two midcourse maneuvers and 90 communications sessions were held with the spacecraft. Similar sessions were held with *Venera 10*. The two spacecraft had trip times of 136 and 133 days, respectively. *Venera 9* arrived at Venus on October 22, 1975, and its lander touched down at 0513 GMT while its orbiter proceeded into an orbit of 69,564 by 936 sm (112,200 by 1510 km) at an inclination of 34°, 10 minutes with a 48 hour 18 minute period. *Venera 10* arrived on October 25, 1975, and its lander touched down at 0517 GMT almost exactly three days later. Its orbiter went into a 70,774 by 1006 sm (113,900 by 1,620 km), inclined 29°, 30 minutes, and having a period of 49 hours 23 minutes.

Both spacecraft landed on the daylight side of Venus; the second and third of the Veneras to do so. The weight of both landers was 3,439 lb (1,560 kg) at separation from their buses. However, the actual weight landed on the planet's surface for the operating spacecraft has not been announced. I would estimate the landed weight, after discarding the heat shield, four parachutes, and related equipment was likely over 2,000 lb (910 kg). Both landers were separated from their respective buses two days before planetfall. Arrival velocity at Venus for *Venera 9* was 35,140 fps (10.71 km/sec) and entry into the Venusian atmosphere, at an entry angle of 20.5° to the horizon, took place at 0358 GMT on October 22. Protected by the heat shield and decelerated aerodynamically, the initial acceleration pulse was 168 g, the velocity was quickly reduced to 820 fps (250 m/s).

At 40 sm above the surface, an extraction chute removed the upper half of the spherical heat shield following pyrotechnic separation of the latter from the lower half of the shield. This initial chute not only removed the upper half of the shield but also unfurled a drag chute, which further reduced the descent velocity to 165 fps (50 m/s). The drag chute acted for 15 seconds and then the three main chutes were deployed; they had a total area of 1940 ft^2 (180 m^2). This occurred at 38 sm (62 km) and 4 seconds later the lower half of the heat shield was discarded.

Broadcasting to the orbiter relaying results of atmospheric analysis commenced at once. After 20 minutes the vehicle had passed through the cloud layer and the main chutes were discarded. The remainder of the descent was guided by a circular, wide hat-brim like structure, which served as an aerodynamic brake for the remainder of the descent through the increasingly dense atmosphere. While no scale has ever been depicted with photos of the landers (except indirectly when a technician was shown in a photo alongside a prototype craft) they may be described as a 6-foot diameter, spheroidal container topped by a truncated high hat whose stiff brim is unusually wide. The cylindrical crown of the hat, perhaps 3 feet in diameter and 2 feet high, acted in a partial capacity as an antenna mount for a helical antenna than can be seen wrapped around the cylinder. At the lower portion of the spheroid are 18 shock struts all equally spaced and attached at their lower ends to a deformable toroid; both struts and toroid intended to absorb the landing energy. Spacecraft overall height appears to be 8 feet. Its touchdown occurred at 23 fps (7 m/s).

During the descent, after the upper hemisphere had been discarded, atmospheric analysis began at once and this data, using a meter band, was transmitted along with engineering data on the condition of the lander and its instrumentation to its orbiting companion. The descent of *Venera 9* took 56 minutes; that of *Venera 10* took 75 minutes. The difference given to the fact that *Venera 10* was in a higher orbit than *Venera 9*; notwithstanding the entry of *Venera 10* at 22.5°, a steeper angle than for its companion. The two spacecraft landed 1240 sm (2,000 km) apart on the daylight side of Venus. Two minutes after landing each spacecraft commenced the broadcast of video, analysis of the surface and engineering data for the lander, and continued to do so even after the orbiting relay stations moved out of range over their respective radio horizons.

While each lander and its relay were in radio sight, the *Venera 9* pair collected and transmitted 53 minutes of such video and an equal amount of surface analysis and spacecraft engineering data. The *Venera 10* pair were in radio contact longer by virtue of the higher apoapsis of the *Venera 10* orbiter, hence 65 minutes of video and analytical data were collected and relayed to Earth. Both landers had searchlights aboard to aid in video acquisition. The lights proved to be unnecessary since the Venus cape was lit by the Sun through the clouds to the extent of appearing, "like a June day in Moscow with slightly overcast skies." That discovery startled the scientists and engineers, since it had been expected that lighting at the surface would be akin to late twilight on Earth, at best. The actual lighting measure was given as 10,000 lux (one lux is the light received on one square meter of surface radially one meter distant from a standard candle).

During the descent, photometers measured light flux in the green, yellow, and red bands and an IR radiometer was active measuring in several areas of the IR spectrum.

Another photometer using an optical spectrum analyzer measured the chemical composition of the atmosphere. Two instruments measured aerosol content of the atmosphere at altitudes of 38-21 sm (63-34 km) and at 38-11 sm (63-18 km). Pressure and temperature were measured from 38 sm (63 km) down to the surface. The aerosol layers were found to be non-uniform with the main layer above 30 sm (49 km). At altitude, the water content of the atmosphere was described as being equal to 0.1 percent that of CO_2.

On the ground anemometers measured ground winds at the *Venera 9* site as 1.31-2.3 fps (0.4-0.7 m/s) while those at the *Venera 10* location were higher at 2.6-4.26 fps (0.8-1.3 m/s). A gamma-ray spectrometer determined that radioactive elements in the surface consisted of 0.3 percent potassium, 0.0002 percent thorium, and 0.0001 percent uranium. A radiation densitometer found the surface to be 2.7 to 3 2.9 gm/cm almost double that found by *Venera 8*, 3 years earlier. The temperature at ground level was measured at 860 degrees F (460 degrees C) and the pressure at 90 atmospheres by *Venera 9*. The *Venera 10* equivalent measures were 869° degrees F (465 degrees C) and 92 atmospheres.

The video pictures at the *Venera 9* site show clumps of well-defined rocks, variously described as geologically fresh, young, and sharp-edged. Those at the *Venera 10* site were described as rather smooth and eroded with rolling clumps in the distance. At both sites very fine dark grained soil filled the spaces between the rocks and crevices. The video panoramas had look angles of 180° by 40° with horizons about 1000 feet away (300 meters) and a field about 650 feet wide at the horizon for *Venera 10* while the *Venera 9* horizon was about 328 feet distant (100 meters). The differences in the two sites imply that some processes are going on that are very disruptive at various planetary locations.[37]

The orbiting *Venera* satellites took photos of cloud cover and the several layers that they form, measured the absorption and reflectivity of the clouds in the IR spectrum, and examined with IR radiometers cloud temperatures in the 8 to 30 micron band. Photometers measured the brightness of the clouds in the ultraviolet band at 3500 angstroms.

---

[37] On the original document, there is a handwritten note about the previous four paragraphs: "Update the page, better data is available now."

Cloud structures were studied and the brightness of reflected Sunlight from the cloud tops was examined. The solar wind was investigated as was its interaction with the Venusian extended atmosphere.

No magnetic field for the planet was found. Cloud top temperature on the dayside was measured and found to be 31 degrees F (-35 degrees C) while the nightside was warmer at -13 degrees F (-25 degrees C). Probing of the ionosphere was conducted and radio occultation experiments performed. *Venera 9* completed its scheduled program in late March 1976. An auxiliary program was added and was completed in April 1976. *Venera 10* was still operating at the end of June 1976 and forwarding data from its orbital aerie to receiving complexes in the Soviet Union.

Thousands of pages of analysis and results will be engendered as a result of the *Venera 9* and *10* investigations. Publication of these are likely to commence in late 1976 and can be expected to continue for some time afterward.

One can expect that spacecraft will be launched again during the 1977 window and it is not too farfetched to surmise that an attempt to bring back a surface sample may be made. Also possible is the chancy business of placing a Venerokhod on the surface with hopes of its survival for hours; perhaps, with luck and good design, for a few days.

For the first time since Soviet planetary launchings began, the Mars window of late 1966, early 1967, passed without any known attempt to launch spacecraft to that planet. Previously Keldysh had made some general remarks that this would occur and these had appeared in print. Few had taken the statements seriously and thus did not appreciate the fact beforehand.

Mars

In February-March 1969 the following Mars window was at hand. The United States had scheduled two flyby Mariner spacecraft and it was expected that the Soviet Union would launch two spacecraft. It did not appear likely that the Soviet Union would forfeit two consecutive launch opportunities. The US spacecraft were launched on February 25 and on March 27. In the March 28, 1969, issue of the *New York Times,* writer Peter Grose, in a lengthy article, stated that an attempt to launch a Mars spacecraft from Tyuratam had failed before the vehicle reached Earth orbit. A failure in the second or third stage of the booster was indicated. (An unidentifiable source has revealed that either the third or the planetary injection stage was used to retrieve the payload.)

A paragraph in Grose's article generated considerable interest: "These difficulties seemed to have been solved in the smaller 2,500-pound spacecraft that the Russians have been using in their exploration of Venus." Grose's comment reinforced the belief in the astronautics community that not only was the Soviet interest in Mars as great as it was for Venus but also that the (new) Mars spacecraft would be substantially larger than the Venera class of spacecraft commencing with the 1969 attempt. It was believed that the Proton (derived) booster was used in the unsuccessful attempt thus supporting thoughts that Mars spacecraft would be very large.

The early Mars attempts used the same booster as the early Venera payloads; as of 1969 this evidently was no longer the case. The workhorse of the Soviet space program was the Vostok booster.

With the addition of a planetary injection stage (i.e., a fourth stage) it served for early launchings in that area and then gradually grew to the Soyuz booster where, among other changes, the Vostok upper (third) stage of 5,440 kg (12,000 lb) thrust was replaced with a 30,390 kg (67,000 lb) thrust stage. A new planetary injection stage was also developed for use with the heavier payloads. The original Vostok booster had a 374,200 kg (825,000 lb) first-stage thrust while with the Soyuz it grew to 635,000 kg (1,400,000 lb). The new booster for use in Mars-bound spacecraft had a first-stage thrust of 1,360,800 kg (3,000,000 lb) and a second stage over 226,800 kg (500,000 lb). The latter two stages constitute the original so-called Proton satellite booster.

For the launching of *Proton 4* a third stage was added thus raising the possible payload in Earth orbit to about 22,680 kg (50,000 lb), however, not all of this capability was used in the *Proton 4* launching. With these three stages and the new planetary injection stage all of the spacecraft to the Moon commencing with *Luna 15* and including *Zond 4, 5, 6, 7,* and *8* were launched. So were all Mars attempts since 1969. A long list of other launchings used this booster including the Salyut space stations, *Cosmos 379, 382, 398, 434,* and almost certainly the *Statsionar* synchronous communications satellite and its related test spacecraft *Molniya 18* and *Cosmos 637*. This large booster is sometimes referred to as the Zond booster despite some possible conflict with the earlier, smaller boosters associated with *Zond 1, 2,* and *3*.

The Soyuz booster as well as its progenitor, the Vostok booster, have either been displayed at the Le Bourget Air Show in France or offered in numerous photographs showing the Soyuz launches. The Soyuz booster was seen, of course, in the Apollo-Soyuz program. The Zond booster, on the other hand, has never been shown, not even so much as in a sketch. Further, there is a belief that its ultimate use has not yet occurred. Until that happens it will, quite likely, be kept from the public eye. Moreover, it will continue to be uprated and this has been evidenced in the *Venera 9* and *10* launchings.

In May 1971, the geometry of the solar system brought Mars and Earth into appropriate juxtaposition for a launch window. Seventy-seven months had passed since the launching of *Zond 2* and no Soviet Mars spacecraft had flown with any measure of success since that November 1964 flight. All the evidence pointed to a multiple attempt and the signs included articles in the Soviet press.

On Monday, May 10, 1971, at 1658.5 GMT a Zond booster lifted off from Tyuratam. It placed a spacecraft in a very low Earth orbit; 174 by 159 km, inclination of 51.4°, and with the short period of 87.7 minutes. However, instead of identifying it as a Mars attempt it received the all-encompassing appellation of *Cosmos 419*. Decay occurred in about 1 day. Of the several possibilities for the failure the most likely cause lies in the planetary injection stage propulsion or guidance systems.

Nine days later on May 19 at 1623 GMT, *Mars 2* was launched from Tyuratam. It had initial parameters similar to those of *Cosmos 419*; however, at 1759 GMT it was successfully placed on a trans-Mars trajectory. A TASS announcement observed that communications with the spacecraft were maintained on a frequency of 928.4 MHz. The Mars explorer weighed in at 4,650 kg (10,251 lb) thus making the capability of the Zond booster clearly apparent. Injection occurred 96 minutes after liftoff; it was the first time that an interplanetary spacecraft from the Soviet Union had made more than one orbit before such injection. Celestial geometry, of course, being the guiding factor.

After another nine days *Mars 3* lifted off on May 28, 1971, at 1526 GMT and injection—not announced—was (presumably) 96 minutes later at about 1702 GMT. *Mars 3* carried instrumentation aboard to study solar radio radiation in the meter wavelength band, which had been designed and built in France under a joint program called Stereo 1. The *Mars 3* weight and broadcast frequency were the same as those given for *Mars 2*. After trip times of 192 and 188 days, respectively, the two spacecraft arrived at Mars at 2019 GMT on November 27, 1971, and on December 2, 1971. Both were placed in areocentric orbits; *Mars 2* at 25,000 by 1380 km, period of 18 hours 4.6 minutes, and inclined at 48°, 54 minutes; and *Mars 3* at 200,000 by 1,500 km, and a period of over 264 hours. Its inclination was only announced as being widely different from that of *Mars 2*.

Both spacecraft carried soft landers but only the one from *Mars 3* gave any indication that it performed although in a very limited manner. *Mars 2* was said to have placed a Soviet emblem on the surface but a fair interpretation of that statement is that maneuvers to place the lander on the planet did occur while the soft-landing mechanism failed in an undisclosed manner. The *Mars 3* lander evidently performed properly until a very few minutes after touchdown. Sketches in the Soviet press show that the lander resembled the *Luna 9* and *13* spacecraft in configuration but were much larger; a reasonable estimate is a factor of 10. That is, the *Mars 3* was 900 to 1,000 kg.

Mars 3 lander touched down, its petal-like shield opened, antennas deployed, and some high-signal, interference-free video ensued for 20 seconds showing nothing. According to TASS no further communications took place. Both spacecraft orbiters returned data for 8 months (an overview is given in *Pravda* of August 25, 1972). The spacecraft sophistication is undeniable and this is well depicted in photos shown in *Aviation Week & Space Technology* of January 3 and 10, 1972, and of May 28, 1973, and in the British journal *Spaceflight* of March 1972. However, the quality control that went into their construction cannot be ascertained. The workmanship that is visible in all of these photos of the full-scale vehicles appears to be excellent.

The fundamental reasons for the failures have not been offered by the Soviet press and thus remain secrets of the Soviet hierarchy. The Soviet effort up to and including the 1971 window, totals 10 tries in which four spacecraft were placed on planetary trajectories; of the latter, two were placed in areocentric orbit and in turn gave rise to two attempted soft landings, which for all practical scientific purposes failed to operate on the surface.

Thus, a huge investment in technical effort and funds has brought a small return in knowledge. The observation has been made that so dismal a record might lead to suspension of the program for several years so that adequate analysis, redesign, and testing could occur and this may have happened. Unquestionably a great deal of activity must have taken place before the 1973 launch window. The key term describing Soviet efforts is perseverance. Their obvious goal was to determine if there is life on the surface of Mars. If positive answers had been found the discovery would have been a quantum leap in both the world of science and throughout the thinking of all mankind. (The *Viking* spacecraft that did land and operate successfully on Mars has left the question of life on Mars unresolved.)

The 1973 Mars window offered the Soviet Union a last opportunity for noncompetitive exploration of that planet since the following window in 1975 would see the US Viking program come to fruition. In view of the desire to be first in answering the question relating to the possibility of life on Mars, it was expected that the 1973 window would be a maximum effort. *Mars 4* was launched at 1931 GMT on Saturday, July 21, 1973; 79 minutes later at 2050 GMT it was placed on a planetary trajectory, which took 204 days to complete. At arrival the orbit insertion retro propulsion system failed and the spacecraft passing Mars at a closest approach distance of 2200 km continued on into heliocentric orbit.

There were some predictions that rather than place a fixed lander on the surface the Soviets would try for a mix of surface rovers together with the expected orbiters. Discussions of a Marsokhod, not unlike the lunar Lunokhod, had appeared in the Soviet press on several occasions. The weight of the latest Lunokhod at 840 kg could easily be accommodated by the 4,650 kg Mars spacecraft. Such a surface vehicle could operate in the immediate vicinity of or right at its landing platform, if necessary, without any locomotion if the surface was too forbidding for movement. Moreover, it could remain buttoned up and protected from bad weather until such time as operation was deemed safe. Deriving electrical power from a radioisotope thermoelectric generator (RTG) and constructed with a low center of gravity it could remain quiescent for as long as necessary.

The use of separate orbiters and landers when not particularly bounded by weight restrictions, as is the case for the Soviet Union, obviates some complex interfaces and permits full commitment of a spacecraft for either the orbiting or lander functions. Calculations made for a *Mars 2* class spacecraft (190-day trip, 4,650 kg), using the arrival velocity taken from planetary tables prepared by Stanley Ross et al., gives rise to 2,410 kg in a 25,000 by 1400 km orbit; more than adequate for first class photography and all the other ancillary missions conducted in and from orbit.[38]

The same weight in orbit could also, in an alternate use, deposit a Marsokhod on the surface, perhaps even larger than the lunar rover emplaced by the Soviet Union. Within four days of the launching of *Mars 4*, the second spacecraft for this window, *Mars 5*, lifted off

---

38 "Study of Interplanetary Transportation Systems—Phase I," S. Ross et al., Lockheed Missiles & Space Company Report, March 17, 1962.

on July 25, 1973, at 1857 GMT; injection occurred at 2015 GMT. Arrival at Mars took place 202 days later on February 12, 1974. The spacecraft was placed in an orbit 32,500 by 1800 km, inclined 35°, and with a period of 25 hours.

On Sunday, August 5, 1973, *Mars 6* was launched at 1746 GMT and injected onto a planetary trajectory at 1905 GMT. The trans-Mars interval was 219 days and arrival occurred on March 12, 1974. An entry vehicle separated from the main bus, entered the Martian atmosphere, decelerated, discarded its heat shield, and then transmitted data for 148 seconds. Then transmission ended and no further data was received. The landing occurred in the vicinity of latitude 24° south, longitude 25° west. The failure was heightened when the *Mars 6* bus retro system did not perform and the bus did not orbit the planet but sped into a heliocentric orbit as had *Mars 4*.

As expected, a fourth spacecraft, *Mars 7*, was launched at 1659 GMT on August 9, 1973, and injected at 1818 GMT. It arrived after a 212-day trip on March 9, 1973. Like *Mars 6* the entry vehicle was separated before planetary arrival but either guidance or propulsion system failure prevented proper operation of the retro system and it missed Mars by 1300 km. Both it and the *Mars 7* bus, which also failed to function properly, joined the *Mars 6* bus in heliocentric orbit.

The performance of these four spacecraft was disastrous; of two intended orbiters only one performed that function, of two intended combination orbiter/landers only one lander performed, but barely. Both the *Mars 6* and *7* buses and the *Mars 7* lander were lost completely.

In discussions with Soviet scientists the economics of interplanetary programs were shown to be no different from those conducted by the United States.[39] It was observed that it cost $66,000 to place a kilogram in Mars orbit for both countries. Elsewhere $110,000 per kilogram was derived as the cost to place that weight on the surface of Mars.

While the four Mars spacecraft weights were never actually announced it seems safe to state that they were at least as large as *Mars 2* and *3* at 4,650 kg each. Thus, of the 18,600 kg en route payload in *Mars 4, 5, 6,* and *7,* some 52 percent could be expected in Mars orbit provided all four vehicles were intended orbiters. Assuming the value for a kilogram in orbit previously noted and using the calculations for weight in orbit discussed above then the total value of the orbiting payload is $640 million. While such calculations are a (good) first approximation some basis for comparison can be made with the Viking program where a total of some 6,350 kg of payload (two combination orbiter/landers) were sent to Mars and where some large fraction of that weight was either orbited or landed.

The program cost is about $900 million so that the en route payload can be costed at $152,000 per kg. If the latter cost is applied to the Soviet 18,600 kg then their program cost is $2.624 billion for the four vehicles, *Mars 4, 5, 6,* and *7.* This, taken to represent costs in the vicinity of what might be termed an upper limit for the program, is obviously a substantial investment, and represents clearly the Soviet interest in exploring Mars.

A reflection on all Soviet attempts to reach and explore Mars, to date, shows 14 attempts with no single total success. And while the partial successes are not to be ignored, they show the existence of some very fundamental design problems or a lack of thorough systems testing together with poor quality control. The Soviet Union has successfully overcome the incredible hostility of the atmosphere

---

39 Private communication from NASA Jet Propulsion Laboratory.

of Venus, which has been amply reflected in the Venera series of spacecraft. Their failures in the exploration of Mars are difficult to understand in the light of their success at Venus even fully realizing that the atmosphere of Mars is at the other end of the spectrum with respect to pressure and temperature.

In examining cost considerations again, suppose the payloads of October 1960, October and November 1962, and November 1964 are (roundly) 907 kg each and the payloads of March 1969, May 1971 (including *Cosmos 419*), July and August 1973 are 4,650 kg each, then based on the value of $66,000 per kg (the cost of orbiting a kilogram) and considering these costs of the program, a total of $2.82 billion is derived.[40] These are really modest costs if comparison is made with the Viking en route-per-kilogram costing. However, it is pointed out that the Viking sophistication is very high while the early Soviet payloads must have been much less so, hence the lower value for program costs may be justified in an overall view. The Zond booster can be priced at least $50 million with its planetary injection stage, the Vostok booster with its planetary stage at $5 million each, which then adds $430 million to the program cost bringing the total thus far to $3.25 billion. Launching services, tracking, and communications via land and expensive-to-operate research ships, data reduction, and other associated costs are certain to bring this figure to over $3.5 billion for the Soviet Union's Mars planetary exploration program.[41]

There is, however, presently no manner in which to determine the value of the broad base that this program has added to Soviet capability in their pursuit of their objectives and goals in space activity. Time is bound to bring success to this program and, with

---

40 The author firmly believes that Soviet intent from the beginning was to land a payload on the Martian surface.

41 Table III offers a brief view of fundamental data on the 14 known launchings.

it, Soviet efforts to extend their goals to the outer planets with emphasis on Jupiter and Saturn.

TABLE III: THE ATTEMPTS OF THE SOVIET UNION TO EXPLORE THE PLANET MARS[42]

Spacecraft	Launch Date Time[43]	Earth Orbit Parameters[44]	Payload Weight[45] (en route)	Mars Orbit Parameters[46]
–	Oct. 10, 1960 –	no orbit	–	–
–	Oct. 14, 1960 –	no orbit	–	–
–	Oct. 24, 1962 1755	217 x 196 65°, 89 min	–	–
Mars 1	Nov. 1, 1962 1614	299 x 200 65°, –	893	–
–	Nov. 4, 1962 1535	158 x 136 65°, 87.7 min	–	–
Zond 2	Nov. 30, 1964 1330	191 x 174 65°, 88.15 min	907[47]	–
–	Mar. 27, 1969 –	no orbit	–	–

---

42 Earth orbit parameters from *NASA Prediction Bulletin*, except as noted below. Where blanks appear, no data was available or event did not occur to yield data.
43 Launch times as GMT.
44 Apogee/Perigee in kilometers, inclination, period (minutes).
45 Trans-Mars payload data from TASS (as published in *Pravda*).
46 Apogee/Perigee in kilometers, inclination, period (minutes). For Mars, of course, it is apoapsis/periapsis.
47 Assumed based on known weight of *Mars 1*.

Spacecraft	Launch Date Time[43]	Earth Orbit Parameters[44]	Payload Weight[45] (en route)	Mars Orbit Parameters[46]
Cosmos 419	May 10, 1971 1659	174 x 159 51.4° 87.7 min	—	—
Mars 2	May 19, 1971 1623	172 x 136 51.5° 87.5 min	4,650	25,000 x 1,400 48.9° 1084.6 min
Mars 3	May 28, 1971 1526	233 x 140 51.57° 88.17 min	4,650	200,000 x 1,500 —, >264 hr
Mars 4	Jul. 21, 1973 1931	203 x 126 51.48° 88.12 min	4,650[48]	—
Mars 5	Jul. 25, 1973 1857	188 x 153 51.55° 87+ min	4,650	32,500 x 1,800 35° 25 hr
Mars 6	Aug. 5, 1973 1746	211 x 154 51.5° 88.18 min	4,650	—
Mars 7	Aug. 9, 1973 1659	208 x 154 51.5° 88.04 min	4,650	—

---

48 *Mars 4, 5, 6,* and *7* weights <u>assumed</u> to be at least as large as *Mars 2* and *3*.

5

Earth Orbit Experiments

As the clouds rise over the Earth, shifting with the winds and changing colors with the position of the Sun so the Cosmos program, initiated on March 16, 1962, varies in its content covering now a reconnaissance satellite, now a scientific foray, upon occasion camouflaging a new mission not yet ready for the public, and when necessary (from the Soviet view) disguising a failure. The word Cosmos is a cover for all missions except those that the Soviets choose to bestow with specific, identifying names. Experiments are conducted aboard reconnaissance spacecraft when it is convenient to do so. Cosmos spacecraft may act as precursors to manned flights and often have in the Soviet space program.

Scientific experiments are conducted aboard spacecraft flying out of all three Soviet launch bases; at Tyuratam near the Aral Sea, at Kapustin Yar east of Volgograd, and at Plesetsk south of Archangel. So far [1975], at least in my view, the major experiments have all come out of Tyuratam—it is the equivalent of the US Cape Kennedy. While Plesetsk and Vandenberg are comparable, Kapustin

Yar is somewhat like Wallops Island except that much more activity occurs at Kapustin Yar by far than at its American counterpart. It would take an encyclopedia to discuss each and every Cosmos flight and some of them could be listed only by number and the orbit parameters that apply so obscure are their missions.

It is probably safe to say that scientific experiments have been aboard most Soviet satellites and I include in this most of the Soviet reconnaissance spacecraft. Of course, many will argue that it is often difficult to separate an experiment from some military application and thus claim it is not a goal in science that is pursued but rather some new weapon or information to aid in the use of a weapon. I will not debate the point for both views are correct; one sees whatever one wishes to see.

Insofar as science conducted in Earth orbit is concerned, the radiation belts and the effects on them due to solar activity have been examined endlessly as have the infinite parades of micrometeorites that pass through the volume of near-Earth space the year-round. The propagation of radio waves through the ionosphere has also been studied at considerable length.

The primary composition of cosmic rays from outside our solar system has also been a topic of prolonged investigation along with the Earth's magnetic field. The early Cosmos spacecraft such as *Cosmos 1* and *2* were used to study the structure of the ionosphere and its latitudinal and longitudinal and time variation. The structure of the atmosphere at satellite altitudes, before the era of satellites largely unknown, and its composition were the object of study by these two satellites.

After the "Starfish" explosion conducted by the US in July 1962,[49] *Cosmos 7*, among other tasks, carried instrumentation to determine the nature of the change in the radiation environment so as to assure themselves that the *Vostok 3* and *4* cosmonauts would not be subject to any harm from that hydrogen bomb detonation. The explosion had in fact increased the radiation count by a factor of seven above the natural space environment.

The *Cosmos 8* spacecraft had piezoelectric micrometeorite sensors aboard and this presumed hazard was examined at orbital altitudes. Investigations showed that this was a very overrated hazard. *Cosmos 9, 10, 12, 13, 15, 16,* and *18* among others measured every conceivable facet of the radiation belts, the effect on them of magnetic storms generated by the Sun, and the lifetime of electrons in the lower belt; the latter turning out to be a function of the Earth's magnetic field strength where for lower strengths the lifetime increased.

The *Cosmos 26* vehicle carried a proton magnetometer, which accurately measured the Earth's magnetic field along the whole length of the spacecraft's orbit.

The first test of television tracking of clouds was done with the *Cosmos 4* craft. Early meteorological spacecraft did not bear the name Meteor as do the present operational birds but carried Cosmos designations. They measured the extent of cloud cover, snow cover,

---

49 ["The Starfish Prime device, with a yield of 1.4 megatons TNT equivalent, was exploded on July 9, 1962, at a very high altitude (approximately 400 km) over Johnston Island in the Pacific, about 700 miles southwest of Hawaii. This exo-atmospheric nuclear explosion released about 1029 energetic fission electrons into the magnetosphere, creating an artificial radiation belt and raising the intensity levels of the natural Van Allen Belt electron population in the inner zone by several orders of magnitude." *The STARFISH Exo-atmospheric, High-altitude Nuclear Weapons Test*, E. G. Stassinopoulos, *NASA/Goddard Space Flight Center, April 22, 2015.*]

and ice fields on both dark and daylight sides of the Earth; dark side coverage was made with the use of IR sensors. Vertical distributions of the temperature, water content, and other atmosphere parameters were made for the lower atmosphere where man pokes around always wondering what is going on overhead.

A good example of a multi-mission spacecraft is *Cosmos 243*, which was primarily a reconnaissance vehicle. It carried, for the first time, equipment to examine the ice coverage in the polar regions and reliably determined the solid ice boundaries around Antarctica during the early part of its extended flight. I was attracted to *Cosmos 243* because it had an inclination of 71.3° and was flown out of Tyuratam the first time (September 23, 1968) that such an inclination had been used from this base. The instrumentation it carried was analogous to the familiar radio telescope except that its antennas were tuned to millimeter and centimeter wavelengths. Centimeter waves are not absorbed by clouds, millimeter waves are absorbed by water droplets, hence the measure of the intensity of these waves as radiated from the Earth's surface gives excellent indication of the cloud cover and moisture content of the lower atmosphere.

Considerable attention was given to making these measurements over the oceans since data collected in these areas is relatively sparse. The oceans are gigantic accumulators, no less, of solar energy and this energy is then reflected in the process of evaporation. The evaporated water then "contains" heat in the form of the latent heat of vaporization; this heat supply is a large contributor to the energy that is found in the cyclones that have so great an effect on the Earth's weather.[50] In taking temperature measurements over the oceans, the data accumulated when examined with other parameters

---

50 Cyclones; the large wind and cloud systems circulating in the atmosphere NOT the storm called by the same name.

that were measured gave a good picture of the weather and its fluctuations over a large part of the Earth's surface.

The density of the Earth's atmosphere at satellite altitudes in the 125 to 190 sm range can be calculated by the effect of the residual atmosphere on the satellite as it passes through perigee on each orbit. Gravitational effects due to irregularities of the Earth do not affect this calculation (they have other effects on a satellite orbit). Since the effect is to lower the perigee and the apogee, and these are measured frequently for all satellites, extensive opportunity was available with the numerous Cosmos launchings to study atmospheric density.

Density was found to be a maximum in the 190 sm attitude range at 1400 to 1600 local time (i.e., at 2 to 4 PM) while minimum density occurred at 0400 to 0600 hours, again local time. During the time of minimum solar activity, the density varied from 60 to 70 percent at the 125 sm range while at the 190 sm range the variation was as much as 200 percent. These density variations were not observed during the years when the Sun was at its maximum activity, a surprising result. All that was necessary for the Soviets to make these measurements was a knowledge of the orientation of the satellite in orbit, its mass (weight) and its dimensions as "seen" by the "airstream" in orbit.

A quantity called the ballistic parameter $CDA/2m$ (a measure of the satellites' resistance to decay from orbit or as some would have it, its vulnerability to decay) is calculated from this knowledge; where CD is a coefficient for air drag on the satellite (2.2 is a good average value for any spacecraft), A is the area of the satellite as seen by the airstream, m is the mass of the satellite (i.e., weight divided by g, Earth's gravity, 32 ft/sec^2). From empirical observations and the use of this parameter a series of curves can be established that permit one to calculate the lifetime of a satellite provided that the orbit is a near-Earth orbit and that its perigee is known at a particular time.

The atmosphere's density was measured by direct techniques by flying instrumentation (ionization manometers) on *Cosmos 108* and *196* in which the amount of ionized gas created by the instruments was directly proportional to the density at the satellite altitude.

On April 18, 1968, the Soviets launched *Cosmos 215* from their Kapustin Yar cosmodrome into a 264 by 162 sm orbit at an inclination of 48.5°. This orbit kept the satellite under the inner Van Allen radiation belt so that the radioactive particles of the belt would not interfere with the satellites recording and other electronic devices. *Cosmos 215* carried eight small telescopes, each of 70-millimeter diameter (about 2.76 inches). Various hot stars were examined from the visible portion of the spectrum to the ultraviolet range.

An additional X-ray telescope was used to study X-ray radiation and photometers were aboard with which to record solar radiation that was scattered in the Earth's upper atmosphere. *Cosmos 215* was an orbiting astronomical observatory (OAO) albeit a short-lived one; it operated in orbit for 45 days. The *Cosmos 262* spacecraft launched on December 26, 1968, also from Kapustin Yar and at a 48.5° inclination, was in effect another small astronomical observatory used to examine the spectrum starting at the visible blue portion and encompassing the ultraviolet and on out to the soft X-ray region.

Ultraviolet in the vacuum of space, the soft X-ray radiation of the Sun, certain hot stars, the interstellar medium and the upper atmosphere of the Earth were examined simultaneously. The ultraviolet and the solar X-rays, the cosmic background, and the Earth's upper atmosphere were examined in 16 selected portions of the spectrum so as to study the quiet regions of the Sun as well as those areas where flares erupted. Radiation from the hot stars and the interstellar medium were studied with the aid of two 10-channel photometers. Additionally, a third set of instruments was concentrated only on the Sun and Sun-tracking sensors were used to activate or deactivate this instrumentation depending on whether or not the satellite was on the daylight or nightside of the Earth. A coordinate system was set up for the satellite so that very reliable determination could be made of the direction from which the recorded radiation came.

Still another spacecraft, *Cosmos 225*, launched from Kapustin Yar was instrumented to determine the quantities and energy levels of electrons trapped in the radiation belt surrounding the Earth at low altitudes; an exercise the Soviets have performed many times.

One way of measuring the energy levels of cosmic rays, especially those from outside the solar system is to measure their deflection in a magnetic field. The Earth's magnetic field strength (about ½ gauss at maximum) is too weak to deflect and therefore, permit measurement of the very high energy cosmic rays. To obtain a magnetic field of the required strength means using enormous magnets and huge amounts of electrical power. This approach gets out of hand quickly and is of little use on Earth in any case since the cosmic rays of interest would strike so many atoms and molecules in penetrating the atmosphere that their energy would be dissipated and scattered over hundreds of square miles making measurement an extremely difficult and expensive task.

If, however, a magnetic field could be created using the phenomena of superconductivity where supercooled metals lose, for all practical purposes, all of their electrical resistance then a magnetic field of great strength is evolved; whereupon its size and weight allow placement aboard an Earth orbiting satellite. The Soviets did this in two satellite experiments; the first on *Cosmos 140* where a 15,000 gauss field was created (30,000 times greater than the Earth's magnetic field strength) and the second on *Cosmos 213* where a 20,000 gauss field was generated.[51] A lengthy paper on the *Cosmos 140* experiment has appeared in a Soviet scientific journal. The two experiments, all of whose results I have not yet seen, were evidently very successful since a plan to fly another such experiment where a magnetic field of 100,000 gauss is to be generated has been made known.

Because the study of high energy cosmic rays is a difficult problem, we know less about them than we would like to. Earth-surface-based experiments are unsatisfactory because of the very indirect "view" we get of cosmic rays and because such experiments are expensive. Soviet scientists realized this, of course, long ago and decided to improve the situation in order to alter this gap in our knowledge.

Whenever the Soviet Union introduces a new launch vehicle into their space program, they include a useful payload on the very first orbital flight. Thus, a success in the booster operation yields a dividend in having a useful payload in orbit. Of course, if the booster fails the payload is lost. Consequently, one would like a choice such that if failure occurs the program concerned with the payload is not wiped out or seriously damaged. On the other hand, some programs for space research receive a low priority because the required payload exceeds the available launch vehicle capability. For new boosters it would also be desirable to have a not overly expensive initial payload; that is, the cost per pound of payload should not be too high.

All of this came together very nicely when the Soviet Union announced the launching of a new spacecraft using a new booster on July 16, 1965; wherein the booster, to this day [1975], has kept the name of its initial payload. The payload was the Proton satellite. The *Proton 1* satellite was announced in an article in *Pravda* as 12.2 metric tons (i.e., 26,896 lb, a metric ton weighing 2,204.6 lb).

---

51 Both of these spacecraft had primary missions involved in the Soyuz program; I discuss this in Chapter 8, "Soyuz."

The launching caused excitement in several quarters. It also was accompanied by a dose of idiocy.

The instrumentation carried aboard the satellite consisted of a huge ionization calorimeter to measure the energy of extremely powerful cosmic rays—up to a 100,000 billion electron volts (10^{14} eV). Compare that number to the electron gun in your color TV set where the energy imparted to an electron in reaching the TV screen is 25,000 eV; the cosmic rays in question are four billion times as powerful as the TV electrons. In a simplification of the Proton equipment, the calorimeter could be described as a multilayered, repetitive sandwich of steel and plastic with photomultiplier tubes mounted on both sides of the repetitive sandwich encompassing all the layers of plastic.

When a very energetic cosmic ray struck the uppermost layer of steel it struck nuclei of iron atoms and gave rise to showers of secondary particles, each now imbued with some of the energy of the original cosmic ray. This first generation of secondary particles then collided with other nuclei in the steel and similar reactions occurred; this procedure gave rise to additional generations of particles until the plastic layer of polyethylene was reached. The collision in the hydrogenous plastic layer caused flashes of light, which were recorded by the photomultiplier tubes.

The higher the initial energy of the virginal cosmic ray the more layers of steel and plastic that were penetrated by succeeding generations of particles. The particles could be traced as they swept through the repetitive sandwich of steel and plastic. Particles due to cosmic ray collisions reaching plastic layer number one but not appearing in plastic layer number two could then be said to have had, as a minimum, a known level of energy since the energy necessary to penetrate the first layers of steel and plastic could be calculated in advance.

This reasoning follows, of course, for higher energy calculations for particles that penetrated additional layers of metal and plastic. The experiment apparatus weighed 22,000 pounds and followed the relatively low cost per pound concept since even especially prepared layers of steel and plastic cost far less than their equivalent weight in electronic instrumentation.

There was other instrumentation aboard, the most important of which was a gamma ray telescope to examine gamma rays at the high energies of 100 million to 1 billion eV. Especially important were the instruments that recorded the direction from whence came the cosmic rays and the gamma rays.

The Soviet Union's scientists were obviously pleased with their new payload but even more so with the new Proton booster, as it was dubbed by the press. In a TASS statement appearing in *Pravda* of August 7, 1965, the booster was described as having more than 60 million horsepower, three times that given for the earlier Vostok/Voskhod spacecraft booster. All other things being equal this would make the first-stage thrust of the Proton launch vehicle equal to three million pounds, a formidable advance in capability for the Soviets. The calculation may not be that simple but even other approaches give answers close to three million pounds of thrust.

If that were, indeed, the correct answer, there were several puzzles to be solved. That thrust was twice the thrust of the American Saturn 1B but the payload was substantially less than that of the Saturn. Part of the puzzle is satisfied by virtue of the fact that the Soviets used storable and not high energy propellants like liquid oxygen and liquid hydrogen; therefore, the energy-per-pound of propellant obtained is considerably less. There has been reference to the use of the storables in numerous places in the "literature" but never any references to cryogenic propellants in their upper stages. However, that answer only partially satisfied the problem.

A regular three-stage booster with the Proton first-stage thrust should have far better capability. *Proton 2* and *3* were launched on November 2, 1965, and on July 6, 1966, from the Tyuratam cosmodrome as had *Proton 1*. They had payloads and orbit parameters very similar to those of the first launch. The dilemma for me persisted until the launch of *Proton 4* on November 16, 1968. The *Proton 4* payload was 17 metric tons (37,478 lb), a 10,580 lb increase over its predecessors; it carried 27,550 lb of instrumentation. That much uprating in a three-stage booster was very unlikely.

After some calculations the answer finally became evident. The original Proton launch vehicle was apparently a two-stage vehicle. The booster for *Proton 4* consisted of three stages. In its three-stage configuration the Proton booster was used to launch the *Zond 5, 6, 7,* and *8* circumlunar flights and a number of other special payloads. Strangely enough there are many publications run by presumably knowledgeable people who right up to the present time have failed to recognize this capability-stretching addition to the original booster. Several publications, I am happy to note, were not long in discovering this important change.

I prefer now to call this launch vehicle the "Zond booster" based on the history of its use following the *Proton 4* launching. I expect that it will see much use in lunar and in interplanetary flights as I noted in Chapter 13, "The Future," on unmanned interplanetary flight and probably will see eventual use for manned flight in Earth orbit.

The *Proton 1* orbital parameters, 390 by 118 sm, at 63.5°, and having a period of 92.45 minutes were close enough to those parameters used in manned flight such that the uninitiated might mistake the intentions implied in the launching. Knowledgeable persons, however, would not have made any such error. In the August 23, 1965, issue of *Aviation Week & Space Technology* one of the uninitiated wrote a letter stating that this satellite really was a manned spacecraft in which six cosmonauts had died. (There is a group of persons who are forever finding cosmonauts lost or dying in space while all of the official facilities of all of the space powers seem never to be able to locate the orbital scenes of these events; strange, huh?)

My own answer to this pathetically anti-Soviet gibberish is best quoted from my letter to that magazine on September 20, 1965.

> On every occasion that the Soviet Union has launched one or more cosmonauts into an Earth orbit, a number of persons lost or in a great fog of ignorance have made unsupported claims about the loss of cosmonauts. The cases presented cover the spectrum from the ridiculous to science fictioneering not suitable for the pulp magazines.
>
> The speculations have become so rampant and the clamor so unceasing that they are now associated with launchings not well understood by most of those who continue to claim cosmonaut losses whether such launchings are directly connected with orbiting men or not. The latest nonsense offered to the public...proposes that *Proton 1* is an orbital sepulcher for six cosmonauts...
>
> The related lifetime (for *Proton 1*) is 220 days; a number quite incommensurate with that of the maximum lifetimes announced for the Vostok and Voskhod spacecraft of 10 and 30 days, respectively...
>
> Furthermore, it is unlikely that the Soviet Union would try to extend flight times for men from 5 days (the then current Soviet record) to 220 days without intervening increments of flight time.

The remainder of the letter was concerned with a technical analysis determining the orbital lifetime and some characteristics of the satellite. The results of Proton experiments, impossible to obtain with any other equipment that had been placed in orbit, are well established with the experiment equipment as described by the Soviets. The history of the flight times of Soviet cosmonauts to date also give the lie to the rantings of the poor misled perpetrators of the tales of orbital catastrophe. So much for the phantasmagorists.

In the strange universe in which we all reside there continually come to light phenomena, which are quite alien to the vast portion of our civilization. One of these concerns the existence of antimatter. Present theories of physics point to the conclusion that for each of the particles known to man in atomic and nuclear physics there exists an antiparticle such that if a "normal" particle as we know it and its anti-image are brought together, they will annihilate each other and the result will be energy in the form of gamma rays. This explanation is a great simplification of a rather complex theory. However, many antiparticles have been discovered, notably the positron—the electron with a positive instead of a negative charge.

The Soviet Academician B. P. Konstantinov has hypothesized that antimatter may constitute certain comets and meteoric matter that enter the solar system, coming from other parts of our galaxy or from other galaxies. He also states that such meteoric bodies may be generated by the separation of these comets. Since the meeting of matter and antimatter will generate more energy per pair of particles (or per unit mass of these) the entry of antimatter meteoric bodies into the Earth's atmosphere should be detectable and very distinct from the entry of ordinary meteoric bodies. Analysis has shown that a spectral line at precisely the 0.511 MeV level (i.e., 511,000 electron volts) should be present if this phenomena occurs.

Cosmos 135, launched on December 12, 1966, from Kapustin Yar, was instrumented for the purpose of investigating this area. Very sophisticated gamma ray detecting apparatus was aboard the satellite. The experiment was conducted during the period from December 13, 1966, to early February 1967. Evidence of the existence of antimatter meteoric bodies appeared to be confirmed since the 0.511 MeV spectral line was observed on two specific occasions and these coincided with the passage of the Geminid and Ursid annual meteor streams. Solid confirmation of these results awaits the completion of additional experiments.

When scientifically oriented satellites are discussed, two pairs of such spacecraft deserve attention front and center for they are landmarks in the early part of the Space Age. At 9:47 GMT on January 30, 1964, a pair of highly instrumented satellites dubbed *Elektron I* and *Elektron II* by the Soviet Union were launched from the cosmodrome at Tyuratam. These were launched into the same orbit plane of 61° in a sequence that set the first into an orbit of 4402 by 252 sm with a period of 169 minutes, while the second went into an orbit 42,400 by 280 sm with a period of 22 hours 40 minutes. This arrangement permitted simultaneous examination of the environment around the Earth at widely dispersed points in the same orbital plane.

Because the phenomena at greatly different altitudes can be very unlike, this investigation was of inestimable value to the scientific community. The orbits overlapped at perigee but differed by almost five Earth radii at apogee. Not the least of the accomplishments in this launching was the fact that the first satellite was separated from the booster while it was still under thrust—i.e., its engine continued to burn during the time *Elektron I* was being placed in its orbital path. The booster then continued to accelerate and placed *Elektron II* in its wide-ranging orbit. TASS was unusually voluble in this case

Chapter 5: Earth Orbit Experiments

as can be observed by the lengthy text of their <u>early</u>—the same day as the launch—release of information.

The total text of that teletype-forwarded information is reproduced here:

* * *

TASS text on Elektron I & II dated Jan 30, 1964/1859 (GMT)

FM SMITHSONIAN OBSERVATORY

TO SUNNYVALE CALIF

3021152

MOSCOW TASS INTERNATIONAL SERVICE IN ENGLISH 1801 30 JAN 64

TEXT MOSCOW--A SPACE SYSTEM--TWO SCIENTIFIC STATIONS/ELEKTRON-1 AND ELEKTRON-2--HAVE BEEN PUT IN ORBIT BY THE USSR TODAY BY ONE POWERFUL CARRIER-ROCKET IN ESSENTIALLY DIFFERENT ARTIFICAL [sic] EARTH SATELLITE ORBITS.

THE FOLLOWING IS THE FULL TEXT OF THE TASS STATEMENT. IN CONFORMITY WITH THE PROGRAM FOR THE INVESTIGATION OF THE UPPER LAYERS OF THE ATMOSPHERE AND OUTER SPACE THE SUCCESSFUL LAUNCHING WAS EFFECTED IN THE SOVIET UNION ON 30 JANUARY 1964 OF A SPACE SYSTEM CONSISTING OF TWO SCIENTIFIC STATIONS—EARTH SATELLITES— ELEKTRON-1 AND ELEKTRON-2 WHICH WERE PUT IN

ESSENTIALLY DIFFERENT ORBITS BY ONE POWERFUL CARRIER ROCKET.

THE SEPARATION OF THE ELEKTRON-1 SPACE STATION FROM THE CARRIER ROCKET WAS EFFECTED ON THE ACTIVE SECTION OF THE FLIGHT WHILE THE ENGINE OF THE LAST STAGE WAS WORKING.

AFTER THE SEPARATION OF THE ELEKTRON-1 SPACE STATION THE LAST STAGE OF THE CARRIER ROCKET CONTINUED ITS FLIGHT ALONG THE PRE-SET TRAJECTORY AND GATHERING THE NECESSARY SPEED PUT THE ELEKTRON-2 SPACE STATION IN A PRESET ORBIT.

ACCORDING TO PRELIMINARY DATA THE STATIONS WERE PLACED INTO ORBITS WHICH ARE CLOSE TO THE CALCULATED ONES. THE PARAMETERS OF THE ORBITS ARE AS FOLLOWS.

ELEKTRON-1–PERIGEE–406 KILOMETERS / APOGEE–7100 KILOMETERS.

ELEKTRON-2–PERIGEE 460 KILOMETERS / APOGEE–68200 KILOMETERS.

THE REVOLUTION PERIODS OF THE TWO STATIONS ARE 2 HOURS 49 MINUTES AND 22 HOURS 40 MINUTES, RESPECTIVELY. THE ANGLE OF THE ORBITAL PLANES OF THE SPACE STATIONS TO THE EQUATORIAL PLANE IS 61 DEGREES.

SCIENTIFIC INSTRUMENTATION / RADIO-TELEMETRIC SYSTEMS / AND RADIO TRANSMITTERS "SIGNAL" AND

"MAYAK" OPERATING ON THE FREQUENCIES OF 19.943 / 19.954 / 20.005 / 30.0075 / and 90.225 MEGACYCLES ARE INSTALLED IN THE SPACE STATIONS.

THE FUNDAMENTAL TASK OF THE LAUNCHING OF THE ELEKTRON-1 AND ELEKTRON-2 SPACE STATIONS IS THE SIMULTANEOUS STUDYING OF THE EARTH'S INTERNAL AND EXTERNAL RADIATION BELTS AND PHYSICAL PHENOMENA CONNECTED WITH THEM. STUDY OF THE EARTH'S RADIATION BELTS BY MEANS OF SPACE STATIONS MAKES IT POSSIBLE TO OBTAIN VALUABLE SCIENTIFIC DATA ON THE NATURE/ DISTRIBUTION IN SPACE/ AND ENERGY SPECTRUM OF CHARGED PARTICLES.

VARIOUS RADIATIONS COMING FROM THE DEPTHS OF OUTER SPACE AND PHYSICAL CONDITIONS IN THE UPPER LAYERS OF THE ATMOSPHERE WILL BE STUDIED SIMULTANEOUSLY.

TRACKING OF THE ELEKTRON-1 AND ELEKTRON-2 SPACE STATIONS AND RECEPTION OF TELEMETRIC DATA IS BEING EFFECTED BY THE COMMAND'S GROUND STATIONS AND MEASURING COMPLEX LOCATED ON SOVIET TERRITORY. THE FUNCTIONING OF BOARD EQUIPMENT IS CONTROLLED BY PROVISIONAL PROGRAM FACILITIES/ INSTALLED ON BOARD THE STATIONS/AND BY SIGNALS FROM THE EARTH.

RADIOTELEMETRIC INFORMATION RECEIVED FROM THE ELEKTRON-1 AND ELEKTRON-2 SPACE STATIONS TESTIFY TO THE NORMAL FUNCTIONING OF ALL THE SYSTEMS.

THE COORDINATION-CALCULATION CENTER IS PROCESSING ALL THE INCOMING INFORMATION.

THE RESULTS OF SCIENTIFIC RESEARCH/ CARRIED OUT WITH THE HELP OF THE ELEKTRON-1 AND ELEKTRON-2 SPACE STATIONS WILL BE PUBLISHED ALONG WITH THE ACCUMULATION AND PROCESSING CF RADIOTELEMETRIC INFORMATION. 30/1859T

* * *

These two satellites were to study in the most all-encompassing manner to date;

- both the inner and outer radiation (Van Allen) belts
- low energy charged particles
- electron and positive ion concentrations
- magnetic fields of the Earth and of the radiation belts
- nuclear components of cosmic radiation
- shortwave solar emissions
- propagation of radio waves of different bands
- galactic radio emissions
- the density of meteoritic matter

It is to be emphasized that both satellites conducted the entire series of experiments, a unique event.

As is usual with Soviet satellites the internal pressure aboard, probably using nitrogen, was somewhat over 1 atmosphere (880-900 mm Hg) and the temperature was a comfortable 65 degrees F. The listing of the actual instrumentation aboard was enough to make a geophysicist not in the program turn green with envy or elated with admiration. These spacecraft were Orbiting Geophysical Observatories (OGO) with the ultimate sophistication of their time.

Again, on July 10, 1964, the Soviet Union launched still another pair of Elektron spacecraft into almost the identical type of orbits that the earlier spacecraft were launched into in order to continue the experiments begun in January.

The Soviets never published any information that I've seen to aid in an estimate of the weight of these satellites. My estimate is that they, together, were some 2,500-3,000 lb while the British publication *Flight Magazine* has published an estimate of 900 lb for *Elektron 3* and 1,200 lb for *Elektron 4*. Presumably, these were equally applicable to the first two satellites also. Finally, while I was at a West Coast tracking station, *Elektron 3* passed overhead; its signal was heard and the station operator estimated that the output power from the spacecraft was between 5 and 10 watts. Quite amazing for that early in the Space Age.

Take a jump now to December 21, 1970, and the launching of *Cosmos 381* into a very circular orbit at about 620 miles inclined 74° to the equator. Like the US-launched Canadian Alouette series, this was a topside sounder, an examiner of the ionosphere. While the Alouettes examined the ionosphere in some half-dozen frequencies, the Soviet bird investigated 20 different frequencies in the 2 to 13.4 MHz band. Additionally, the emission from the Sun in the ultraviolet was also intensely examined. The solar emissions were examined between 3 and 1,500 angstroms. As usual, various aspects of the radiation environment were also investigated.

Because the spacecraft was not always in radio sight of the Soviet Union, a large capacity memory system was part of the instrumentation; together with sensors to determine the orientation of the Sun with respect to the spacecraft. These recordings were both clued with preset/programmed timing marks—so that the ionospheric and other recorded results not only were played back when the spacecraft was over the Soviet Union but also that the spacecraft location and orientation could be determined as to better determine a "map" of all of the recorded information. An onboard magnetometer aided in this mapping process.

Despite the fact that the ionosphere is a dynamic mechanism, always changing, enough information was obtained to statistically form a spherical map to determine the global characteristics of the ionosphere. When one reflects on the fact that it is these characteristics that determine the nature of most radio communications activity on Earth the importance of the topside sounder *Cosmos 381* is made readily apparent.

There must have been much soul-searching among the secrecy-prone managers and military persons who direct the Soviet space program before they decided to take advantage of the good that comes with international cooperation. Even then, the crack in the door was distinctly limited. Such cooperation, heralded long and widely, was limited to members of the Eastern Bloc. *Cosmos 261* was launched on December 19, 1968, at a fraction of a minute before midnight, GMT, from Plesetsk complex. It went into an orbit 135 by 416 sm, at an angle of 71°, and with a period of 93.1 minutes. The radiation characteristics of the lower space environment were once again the subject of study along with the aurora at the northern latitudes—the aurora borealis. Involved were "The People's Republic of Bulgaria, the Hungarian People's Republic, the German Democratic Republic, the Polish People's Republic, the Socialist Republic of Romania, the Soviet Union, and the Czechoslovak Socialist Republic."[52]

---

52 No reference given in original document for this quoted text.

A large number of scientific papers resulted from this series of investigations; there were as many as 15 coauthors for single papers. The shades of secrecy lifted still another measure in these papers, for much of the instrumentation aboard *Cosmos 261* was described in elaborate detail including drawings and explanations of the working of the instruments—good! As has occurred in a number of instances with all the space powers, the *Cosmos 261* rocket stage (the last stage of the booster always goes into orbit with the payload unless steps are taken to deorbit it immediately after payload separation) broke up and left almost two dozen pieces of debris in orbit. They have long since decayed as has the satellite itself.

Having initiated this program, the Soviets were to follow it up with a long series of additional spacecraft and then formally named the Intercosmos series. All have been used for the same sort of investigations although not all the bloc countries participated in all of the satellite experiments. There is one difference worth noting; *Cosmos 261*—which can reasonably be called *Intercosmos 0*—was launched from Plesetsk launch complex.

InterCosmos 1 through *5*, have been launched from the Kapustin Yar launch base. As a matter of fact, *Intercosmos 5* was the only launch from Kapustin Yar during the entire year of 1971. Then through 1975 there were nine more Intercosmos launched.

Biological Experiments

Scientific experiments, almost no matter how interesting, usually attract the community of science rather exclusively. The public at large, at least as I see it, has a peripheral interest at best. But place living creatures aboard a spacecraft and interest perks up all over the place, like quills on a porcupine. When the creatures turn out to be "man's best friend," the interest grows larger. So, on [US President George] Washington's birthday no less, February 22, 1966, *Cosmos 110* was launched into a 116 by 562 sm orbit, at an angle of 51°54', with a period of 95.3 minutes. Aboard were two dogs, Veterok (light breeze) and Ugolyok (ember). What surprised me was the apogee altitude, the highest for a living creature since the beginning of the Vostok precursors carrying dogs. It seemed evident to me that here was the precursor to a new manned program and the altitude spoke of long-lived spacecraft, SPACE STATIONS.

This excited me to a great extent, for I established my reputation in the astronautics community on the basis of my original work in space station design—the first realistic engineering study of such vehicles was conceived, initiated, and directed by me.

The first design for a space station on which a patent was awarded in the US was also completed by me, so was its conception and initiation. The NASA space tug and the Space Shuttle are both concepts conceived and presented by me in two papers before learned societies in 1958 and 1960, long before they showed up in

NASA programs. Mine were much more modest designs and are long since outdated in most aspects by the NASA spacecraft.

The highest manned flight prior to this time was about 307 sm. Here we were more than 250 miles higher.

The two dogs were in separate compartments aboard the satellite and were instrumented with many sensors examining all the bodily biological functions. The cardiovascular functions were prominent among these; probes were implanted in the heart and in several arteries. Studies were made of the animals' response to weightlessness and to the presence of protons from the lower edge of the inner Van Allen belt. In addition to the several dozen biological studies, the dogs were studied via television from control centers on Earth. The usual coterie of lesser experiments was also aboard—bacteria strains, chlorella, yeast, blood serum, and so on. Dosimeters and emulsion blocks recorded all radiation levels for comparison with the various effects on the dogs (if any).

The compartments were well ventilated and waste was collected and removed from the compartments. The dogs were fed through tubes directly implanted in their stomachs. The Soviets, it should be recognized, are utterly expert in animal experiments using dogs and have a long history of successful experimentation using these animals. Veterok and Ugolyok were brought back to Earth, recovered safely, and displayed to the public two days after they landed on March 16. The dogs had spent 522 hours in space, almost 22 days, a record that would stand until the manned Salyut flight of 1971.

Very clear photos in the *New York Times* of March 19, 1966, show the two dogs still wearing part of their space instrumentation and looking well fed and healthy, which was the description given by Dr. Boris Yegorov, a member of the three-man *Voskhod 2* flight in October 1964. Only much later, at the 17th Congress of the International Astronautical Federation in Madrid in October 1966, did Soviet scientists reveal that the dogs had lost up to one-third of their preflight weight during their stay in space. A considerable part of the coordination was also lost, but recovered quickly within a few days after returning to Earth. Movie film shown of the dogs four months after their return showed them romping about as would any normal playful canines. Thus, they joined a very exclusive club of canine cosmonauts—of which there are much fewer than their human counterparts.

Up to the close of 1975, mammals, such as dogs, have not again appeared on the scene in space following the orbital sojourn of Veterok and Ugolyok. There have, however, been five Soviet flights that were known to have biological specimens aboard. (The Zond circumlunar flights, which carried biological specimens are not included in these discussions of Earth-orbiting vehicles.) At least one of these was not identified as carrying such specimens until long after the flight. This was *Cosmos 368* launched on October 8, 1970. It did not attract great attention even then because, it appears, that only lower forms of life were carried; such as yeast cells, various seeds, and onion bulbs. If higher forms of life were aboard, this has not been made clear. The specimens were aboard for the purpose of testing in a radiation environment under conditions of zero gravity. They were conducted successfully. Following a 6-day flight, the vehicle was recovered after a soft landing.

Again, there was a long hiatus for "biological" spacecraft lasting until October 31, 1973, when *Cosmos 605* was launched from Plesetsk into a 263 by 137 sm orbit, inclined 62.8°, with a period of 90.7 minutes. This spacecraft literally carried a zoo full of animals; the number was so large because the scientists conducting the experiments wanted to attain statistically reliable results when evaluating those experiments. Several forms of plant life were also carried aboard. The tests were conducted to aid not only in the determination of the long-term effects of spaceflight on man but also for effects related to his support during such intervals. The TASS release stated that "during the flight, the influence of spaceflight factors on living organisms will be investigated further, and tests will be conducted on the life support systems of the biological objects."

Aboard *Cosmos 605* were several dozen white rats, six boxes of tortoises, flour beetles, a mushroom bed, drosophila fruit flies, and various cultures of bacteria. A control group was maintained in a ground laboratory for the duration of the flight. All of the methods for life support were identical for the orbital and ground specimens. For the first time a second generation of insects was obtained during a spaceflight, which aided greatly in predicting long-term effects and the possible transmittal of any genetic mutations that might occur due to radiation or weightlessness or to a synergistic combination of these conditions.

An experiment of considerable interest that was conducted aboard the vehicle was the introduction of an electrostatic field around the spacecraft as a technique to deflect charged particles from the craft. This was evidently pointed at long-term interplanetary voyages for humans as well as for their protection aboard Earth orbiting space stations during solar radiation storms.

Detailed pathological studies were performed on the animals during the weeks following the recovery on November 22, after 22 days of spaceflight. The results were published in Soviet journals on spaceflight and in medical journals.

One year later on October 22, 1974, *Cosmos 690* carrying a group of biological specimens similar to those previously onboard *Cosmos 605* was launched from Plesetsk. This spacecraft had orbit parameters very much like those of its predecessor. It was, in effect, a repeat of the *Cosmos 605* mission even to the statement issued by TASS describing the purposes of the flight. Twenty-one days later on November 12, the spacecraft was recovered and, again, follow-up pathology and other related examinations were conducted to ascertain the space-related effects on the 40 white rats and other specimens that had been aboard.

For some time, the US and the Soviet Union had carried out detailed discussions on the possibilities and requirements for conducting mutual spaceflight oriented biological research. These lengthy and delicate efforts came to fruition in late 1975 when *Cosmos 782* was placed in orbit from the Plesetsk complex on November 25. The orbit, again, for all practical purposes, was the same as that of prior biological research satellites *Cosmos 605* and *690*. Almost a metric ton (2,205 lb) of instrumentation was carried aboard for this research and this included a centrifuge with a diameter of 26 inches (66 cm). Contributions to the onboard experiment packages were made not only by the Soviet Union and the US but also by Czechoslovakia and France.

Participation in the research effort on this multinational team also included scientists from Poland, Hungary, and Italy. The white rats used in the experiments had been delivered by Caesarian section in order to keep them almost germ-free. While the rats were kept on a stable platform in a zero-gravity environment, the tortoises aboard were divided up to be placed in a partial (0.6) gravity environment and in zero g so that comparative pathology could ensue after the flight.

The spectrum of biological specimens aboard was again similar to those carried aboard *Cosmos 605* and *690*. There were also numerous experiments carried out in the area of radiation physics directed toward the protection of man in future spaceflights. On December

15, 1975, the spacecraft was returned to Earth in the vicinity of Karaganda in the Soviet Union whereupon the scientists of the seven nations involved plunged into analyses of the results of their orbital efforts.

More or less in parallel with this biological research study the unmanned *Soyuz 20* spacecraft had been launched from Tyuratam on November 17, eight days before the *Cosmos 782* was placed in orbit. Two days later, *Soyuz 20* docked with the space station *Salyut 4*. On December 5 the Soviet Union announced through TASS that, "onboard the *Soyuz 20* complex biological experiments are being conducted simultaneously on various plants and living organisms." Twelve kinds of plants including gladioli, cacti, and beans were aboard along with turtles and the ever-present drosophila flies.

Statements were made that a number of these *Soyuz 20* experiments were analogous to those aboard *Cosmos 782*, but that the microclimate aboard the two vehicles differed. The biological aspect of *Soyuz 20* had not been announced beforehand. *Soyuz 20* stayed in orbit with the *Salyut* station for 91 days. Details of those experimental results were not generally available. *Soyuz 20* was safely recovered on February 16, 1976.

Very much more complex biological/medical research efforts can be expected to be performed by the Soviet Union. However, these are likely to be conducted onboard the *Salyut* stations rather than the space-limited Soyuz spacecraft. Such experiments are certain to require sophisticated equipment and the presence of highly trained personnel—research doctors of the medical profession—rather than cosmonauts given special training for such purposes. The next few years should see such events occur periodically.

6

To The Moon— Soft Landing or Bust

No single effort by the Soviet Union shows more persistence than their efforts to make a soft landing on the Moon and to be the first to do so. Their eventual success in February 1966 came only after a long string of failures, mishaps, and near successes.

The first three lunar missions have been well enough documented and discussed in Chapter 1, "Beginnings"; they were not attempts at a soft landing. The first serious attempt to make the coveted soft-landing prime mission came on January 4, 1963, but the translunar injection stage failed to ignite and the spacecraft lasted for a very short time in Earth orbit and then decayed. On February 3, 1963, a second attempt was made, but the third stage of the booster-to-Earth-orbit malfunctioned and the spacecraft remained in orbit for a very short while; the spacecraft reentered the atmosphere and was destroyed.

On April 2, 1963, *Luna 4* achieved a translunar trajectory and was announced by the Soviet Union as a "lunar spacecraft" but with vague remarks as to purposes and intents of the mission. No wonder, after a few days of flight it became apparent that the spacecraft was going to pass behind the Moon since its velocity was insufficient to place it at the proper location for landing. On April 6, the spacecraft passed the trailing edge of the Moon at a distance of 5,300 miles. A few follow-up articles appeared in the Soviet press about the mission and then all was silent.

On March 12, 1965, at 0920 GMT the spacecraft *Cosmos 60* was launched into Earth orbit. Examination of the launch time, one day past first quarter (phase of the Moon), and the orbital parameters established the fact that this vehicle decayed after a few days in orbit, with no real discussion that its purpose was intended to "go" to the Moon. Somewhat different had been *Cosmos 21*, which was launched on November 11, 1963, about three days before last quarter; its launch-lunar-phase relationship matches the later launches of the *Zond 5*, *6*, and *8* circumlunar spacecraft. *Cosmos 21* (like *Cosmos 60*) had a short orbit life and decayed, after 46 orbits, on November 14, 1963.

At 0753 GMT, on the morning of Sunday, May 9, 1965, the drive to the Moon began with a renewed level of interest as *Luna 5* was launched from the Tyuratam cosmodrome. There was an air of confidence exuded by the Soviets on this occasion and the rumors flew thick and fast on the matter of just what type of lunar vehicle was aboard *Luna 5*. This 3,254 lb spacecraft was 119 lb heavier than *Luna 4*. The balloon of confidence burst, however, when after the command to retro for the soft landing was given, a malfunction in the program led to a too-early retro firing and the spacecraft landed too hard to survive.

Chapter 6: To The Moon—Soft Landing or Bust

On June 8, 1965, *Luna 6* was launched into a translunar trajectory and the Soviet Union announced its flight plan as being three-and-a-half days to the Moon. It weighed 3,179 lb; this was the only time in the history of the early Luna vehicles (up to and including *Luna 14*) that the spacecraft weight was lower than its immediate predecessor. A midcourse correction was scheduled for late in the day on June 9. The engines came on as scheduled but they failed to turn off. As a result, the spacecraft acquired an additional velocity of some 7,700 fps and thereupon missed the Moon by 100,000 miles. A normal correction would have added perhaps 100-150 fps to the spacecraft velocity.

The size of the velocity increase yielded a clue as to the nature of the engine used for the midcourse corrections on the *Luna* spacecraft and revealed an interesting characteristic of its propulsion system. The fact that an excessively burning engine led to so large a velocity addition clearly indicated that the propellant source for the main engine was totally available for the midcourse correction. Thus, the evidence showed that midcourse engine and the retro-engine were one and the same; with a single propellant source for both functions. A good estimate of the size of this engine is given later in this chapter along with the pieces of the puzzle that helped me to make the estimate.

Bit by bit the shards of camouflage that secreted the Soviet lunar program were being removed both by the Soviets and by my own analysis. On the eighth anniversary of the Space Age, October 4, 1965, the 77th launching of 1965 occurred at 0737 GMT and *Luna 7* was on its way to the Moon. Once again, a midcourse correction was made—on the morning of October 5—and once again the spacecraft weight showed an increase, this time to 3,320 lb, and once again the retro program failed to function properly with the unhappy result of an impact on the Moon hard enough to damage the spacecraft beyond a working condition. Impact took place at

2208:24 GMT on October 7. There were some published rumors that *Luna 7* had impacted at 66 fps instead of its intended 27 fps; no confirmation of this information was ever made available to me, but were there any truth in this, then a soft landing was not far off. One more attempt should bring the necessary refinement to affect a successful landing.

The wait was not long for on December 3, 1965, along came *Luna 8* from Tyuratam at 1043 GMT. I thought to myself that this should do it; and it would be well before the soft-landing attempt by the American *Surveyor* spacecraft. As expected, *Luna 8* was heavier than its predecessors at 3,421 lb. It also offered a new surprise; gone was the 65° inclination of the earlier Luna spacecraft. *Luna 8* was launched at the inclination of 51° used in November 1965 in conjunction with the *Venera 2* and *3* launchings. At 1900 GMT on December 4 a midcourse correction was made. Numerous communications sessions were held with the spacecraft during its translunar trip.

I waited with great anticipation hoping that the long strived-for success was at hand, but once more it was not to be. I listened to the reports on December 7 telling of yet one more failure; *Luna 8* had hard-landed at latitude 9°8' north and longitude 63°18' west in the Oceanus Procellarum at 2151 GMT on December 6.

I walked about muttering to myself for the next several days wondering what the devil could be going wrong and, arriving at some conclusions, promptly fell into a trap borne out of lack of confidence in my own opinions and the enchantment of pretty sounding nonsense proposed by others. In a letter to the editor of the now defunct magazine *Missiles & Rockets*, I suggested that the lunar surface consisted of a low-density structure, easily pulverized by the landing spacecraft, thus burying them. I had previously been an adherent of the solid surface theory, which later proved to be

Chapter 6: To The Moon—Soft Landing or Bust

quite correct. Some very eminent names had proposed the highly vesicular, low-density surface; but let my letter tell this.

From *Missiles & Rockets*, December 20, 1965

> To the Editor:
>
> The current failure of the Soviet spacecraft *Luna 8* to complete a soft landing on the surface of the Moon following three previous attempts (*Luna 5, 6, and 7*) warrants immediate and specific attention...
>
> We cannot but agree that the Soviet lunar spacecraft must be quite sophisticated...What, then, went wrong? I would like to speculate that perhaps little or nothing with the Soviet spacecraft. The fault appears to lie in our own distinct lack of knowledge of the lunar surface...I should like to hazard the guess that what we have seen in the Ranger photographs is the pulverized surface of the "fairy castle" (cotton candy) theory suggested in recent literature. If, indeed, the upper few inches of the fairy castle structure has been pulverized by meteoroid strikes and underlying this is several yards of this very-low-density structure, then we might expect that a vehicle attempting a soft landing would, no matter how accurate its instrumentation and propulsion systems, cause the surface to fail and then be covered by debris to a depth of several yards. The dielectric constant of the debris would prevent even battery-powered communications from taking place since the effective wavelength radiated by any antennas aboard would be distorted beyond usefulness....

Well, it sounds nice but turned out to be sophisticated bunk. As late as the landings of *Apollo 14* and *15* and *Luna 16* and *Lunokhod 1*, no such low-density surface structure had been discovered on the lunar surface.

In a letter to lunar scientists in the Soviet Union, I proposed the possibility of the fairy castle structure. I also proposed that a 5-meter telescoping rod be extended from the Luna spacecraft during the landing process so that the end of the rod containing a microswitch could shut off the propulsion system and so ensure that the vehicle would not lift off accidentally after touching down in landing. The secrecy syndrome that has been so much a part of the Soviet space program prevented any answer to my letter, copies of which were sent to *Pravda*, the Soviet equivalent of *Aviation Week* (*Aviatsiya and Kosmonautika*), to L. Sedov, V. S. Troitskii, and to N. Barabashev—the latter two are well known astronomers and lunar scientists. This tale has an interesting if coincidental ending, about which more shortly.

The morning of January 31, 1966, started out as many Mondays do, a bit of a bore and with fading memories of a pleasant weekend. The Soviet Union peeled off the dullness of the day with the announcement that they had once more launched a spacecraft to the Moon; *Luna 9* had been launched at 1053 GMT. It was 69 lb heavier than *Luna 8* and weighed in at 3,490 lb. It was launched into a typical parking orbit of 139.2 by 107.5 sm, at an inclination of not quite 52°, and having a period of 88.5 minutes. A midcourse correction of 71.2 meters/sec (233.54 fps) was made on February 1 at 1929 GMT.

I usually lunch with one or more colleagues, but just before noon on Thursday, February 3, I found myself driving off to a restaurant alone, listening to the radio and musing about nothing in particular. The noon news broadcast came on and the first thing I heard was the bland but puzzled voice of a local reporter saying that the Soviet Union had landed a spacecraft on the Moon. Puzzled voice continues...the landing was a soft landing and the spacecraft was *Luna 9*. I let out a piercing shriek, flipped the dial to another station, and listened to an excited voice repeat, but with understanding, the data on the landing. At last! I ate lunch quietly but quickly and headed back to my office. The place was abuzz with commentary, speculation on the near future, dismay at our (US) now second place in the soft-landing effort, and liberally sprinkled with sour grapes / disparaging remarks from a few far-right-of-center nuts that are found everywhere.

The flight of *Luna 9* had taken 79 hours 52 minutes from liftoff to landing. The landing occurred at 1845:30 GMT on February 3 and was located in the Oceanus Procellarum at latitude 7°8' north and longitude 64°22' west. The retro-engine was ignited just 48 seconds before the landing at 1844:42 GMT at an altitude of 75 kilometers. Before retro, all mechanisms not used in the landing and not connected with the surface payload were ejected thus lightening the load for the retro-engine. The landed payload weighed 220 lb, a rather low efficiency it seems for the starting weight in translunar trajectory of almost 3,500 lb. On the other hand, all of the auxiliary structure, controls, etc., were also brought to the Moon but discarded in the landing process. As a consequence, it is difficult to compare the landed weight of the (later) American Surveyor spacecraft with the Luna since the former discards only the casing of its solid propellant rocket but keeps all other structure intact, through and after the landing process. In essence the Soviets separate their instrumented payload whereas the Surveyor keeps all of its payload and other mechanisms with no such separation.

Four minutes and 10 seconds after landing, the antennas on *Luna 9* unfolded and the first radio session began. The photos of the lunar surface are well known, as is the incident in which Sir Lovell of Jodrell Bank, England, received them and released them to the press while using an incorrect conversion scale so that the horizontal was compressed thus giving an impression of a much rougher lunar surface than existed in fact. The Russians expressed their annoyance at this, justifiably so. The next time he obtained photos broadcast from the Moon he asked permission for their release—and received it—and for the correct scale—and got that too.

Transmission of the video photos was uncoded—no unnecessary computer derived plan to secret the photos. A regular newspaper-type facsimile receiver such as that used to receive wirephotos sent by Associated Press (AP) or United Press International (UPI) was put to use by the Soviets and by Sir Bernard Lovell as well.

Luna 9 was battery powered and the power available for data broadcasting exceeded that originally calculated so that one more communication than was planned for was held commencing at 2037 GMT on February 6. A total of seven radio-communications sessions were held with *Luna 9* for a total broadcast time of 8 hours 5 minutes.

The results of this historic event were widely published. They included a substantial addition to the scientific and engineering knowledge of the Moon. The landing kicked into the dust the fairy castle structure proposed for the upper lunar surface. It was never again discussed with any enthusiasm and never again accepted as a credible theory.

The hardness of the lunar surface and the distribution of craters down to micro sizes as discovered by *Luna 9* is now well established by the multiple visits of Soviet and American spacecraft to the Moon. History, however, has written that the first lunar soft landing was made by a Soviet spacecraft and the firmness of the surface therein confirmed for future landings—a most important fact for the drive into space. Moreover, safety for the Apollo landing was assured. A conference on controls and automation was held in Vienna in 1967.

At that session papers were presented on the controls of *Luna 9* during the landing procedures. One portion of the landing system called for and used a 5-meter rod that unfolded before the landing; the rod had a microswitch at its tip, which shut off the rocket engines so as to prevent any disturbance of the surface or possible liftoff of the spacecraft after landing. Coincidence? Who knows? I've often mused about that rod and my letter but I don't expect to ever get that one resolved.

Filled now with the confidence of success the Soviets decided to lose no time and so promptly at the next opportunity on March 1, a Tuesday, another lunar launch was initiated. *Cosmos 111* launched at 1103 GMT that morning failed because its translunar injection stage did not ignite. It soon decayed and was lost in the debris of space history. Since lunar launch opportunities are 28 days apart and the launch window lasts for about 4 days, March offered still another crack at the Moon.

Not one to let such moments pass the Soviet Union launched *Luna 10* on Thursday, March 31, at 1047 GMT. The parking orbit was similar to that of *Luna 9* and once again the weight increased, this time to 3,527 lb. The Soviets stated at once that their intent was to create an artificial satellite of the Moon. So, it was not to be another soft landing. The translunar trip took 79 hours 57 minutes from liftoff to injection into selenocentric orbit at 1844 GMT on Sunday, April 3. By its 14th orbit the spacecraft had been set into an orbit of 631.9 by 217.5 sm, at an inclination of 71°54', and with a period of 178.25 minutes. Recall that for the same altitude a satellite in orbit around the Moon has a longer period than it would at Earth because of the lesser mass and smaller radius of the Moon.

The satellite separated into two parts, one of which was a complete scientific package that weighed 540 lb. The instrumentation aboard was used for studies in micrometeorite flux and energy, solar plasma, the lunar magnetic field (if any), gamma and X-ray radiation from the lunar surface, solar radiation, and the lunar gravity field. During the active life of *Luna 10* (i.e., as long as it had electrical power) 219 communications sessions were conducted with the probe. It had completed 460 orbits when the last communications session was held on May 30, 1966, and had traveled the equivalent of over 4,300,000 miles up to that time.

The first lunar satellite gathered enough data to keep several teams of scientists busy for months. Some idea of the data collected can be obtained from the following quote from *Pravda* of June 3, 1966:

> Nine gamma-radiation spectra of the lunar surface and ten magnetographic sections of the near-Moon space in the 352-1,016 kilometer (218 x 630 sm) altitude range were obtained during the time the (active) probe was in near-lunar orbit.
>
> 74 trajectory measurement sessions, 17 prolonged sessions for measurement of radiation conditions, properties of the near-moon plasma, density of micrometeorite material, thermal radiation of the moon, and three sessions for measurement of the x-ray fluorescent radiation of the lunar surface were conducted during this period. The trajectory measurements were utilized to analyze the evolution of the *Luna 10's* orbit and to estimate the anomaly of the Moon's gravitational field.
>
> A large number of brief sessions to check scientific measurements (probably calibrations) and one session of the radio-setting (properties) of the moon were also accomplished.

Some of the results obtained could be given after rather preliminary examination of the data. The same *Pravda* article gave the following information:

> The magnetic field in near-Moon space is very weak and its intensity fluctuates between 17 and 35 gamma on different days; (by comparison the Earth's field is 50,000 gamma at maximum). The gamma-ray spectra of the lunar surface are similar in nature to the gamma-ray spectra of (Earth's) basalt rock;[53]
>
> The spatial density of micrometeorites in near-Moon space is higher than in interplanetary space;
>
> An elevated low-energy particle flux density, particularly of electrons, is observed in the vicinity of the Moon;
>
> The anomalies of the moon's gravitational field are slight. (For close-in orbits that Apollo used this turned out not to be true. Large mass concentrations, "mascons," buried beneath the lunar surface affected spacecraft orbits a great deal meaning therein that the gravity field varied considerably.)

On its arrival at the Moon, on April 12 (Cosmonautics Day), and on May 1 the Soviet anthem "Internationale" was transmitted from *Luna 10* to various audiences in the Soviet Union. The Soviets were not about to miss any opportunity to let the world know that their spacecraft had been first on and around the Moon. First again!

The next flight to the Moon did not occur for several months. A clue to some change in the lunar program was to be found in the

---

53 There is no connection between the gamma ray and a gamma of magnetic flux. The names are merely a coincidence.

launching of *Luna 11* at 0805 GMT on the morning of August 24, 1966. As usual the spacecraft weight was up, this time to 3,616 lb. The flight time to the Moon increased to 85 hours and 43 minutes; the payload had been increased and the launch velocity decreased. One can trade off the flight time and velocity against payload for some total fixed spacecraft weight; of course, there is a lower limit for the velocity that must be reached or exceeded in order to reach the Moon in the first place. The Soviets would not brush up against that limit until the flight of *Luna 18* in September 1971. There was, in 1966, a fair amount of room for trading off these variables against each other in this realm of what the future would term "small spacecraft."

The communiques on *Luna 11* were few. A description of its orbit around the Moon—it came within 99.4 miles of the surface and had an apolune at 745 miles; its inclination was 27° and its period was 2 hours and 58 minutes. It would appear that *Luna 11* was an intended photographic mission, which failed. Some stories appeared in the press that in fact some photos of poor quality were obtained and other tales that the spacecraft attitude control was lost—that is, the spacecraft tumbled in its orbit. In any case, after 277 orbits of the Moon, *Luna 11* ran out of power on October 1, 1966, and the mission was concluded. No doubt even at this writing it is still orbiting the Moon.

Luna 12 was essentially a duplicate of the *Luna 11* mission, except that it worked and returned photos that were published in *Pravda* and other Soviet newspapers. It was launched on the morning of October 22 at 0842 GMT. It ended up in a lunar orbit with perilune of 62 miles and apolune of 1079 miles. What wasn't known for a long time; until the *Luna 17* mission had deposited *Lunokhod 1* on the surface, in fact, was that *Luna 12* also carried copies of the electric motors that were to run the Lunokhod wheels. They were tested in the cold of deep space where they operated successfully, thus providing another step in what was to become one of the most successful lunar missions ever as *Luna 17* must be reported in the annals of history.

Because *Luna 12* continued to transmit until January 19, 1967, by which time it had completed 602 orbits, one can better understand the original intent of *Luna 11* and the gap by which it missed its intended mission. The blow was softened somewhat by virtue of the fact that both *Luna 11* and *12* collected extensive information on the (soft) X-ray radiation from the lunar surface and from deep space far beyond the Moon. The encyclopedia of knowledge needed for man's eventual penetration into interplanetary space was being compiled bit by bit.

I thought it strange that the Soviets had not followed up their first soft landing with another soon after the first success. At one point I reflected on the possibility that the failure of *Cosmos 111*, launched in early March 1966, may have been a soft-landing attempt but, of course there is no way of ever knowing until the Soviets decide to discuss the topic. It could just as easily have been the intended first lunar orbiter for them. There is still no accurate way of forecasting their intentions. Only the forecastable geometry of the solar system tells us what is possible at any given time, not what it is that they are likely to do or when.

Well, it was a dull time of the year when the Soviet launch bases were deep in frigid grasp of winter where temperatures 30 or 40 below zero are common. So I was pleased and surprised to hear that *Luna 13* had been launched in the early hours (early for the US) of Wednesday, December 21. Ah! I said, "a landing for Christmas." Sure enough; 80 hours later at 1801 GMT the spacecraft landed in the Oceanus Procellarum latitude 18°52' north and longitude 62°3' west. This is the left side of the Moon as one looks up at it, not far above its equatorial belt. This landing area was not far from the site of *Luna 9* (about 250 mi).

The two spacecraft were much the same in appearance and although weights for *Luna 13* were never announced, I have assumed that there was little or no difference. A boom was extended from *Luna 13* that had a charge of explosive, which was used to drive a rod into the surface some 30 centimeters (about 11.5 inches). The rod was titanium tipped and was driven with a force of about 15 pounds. The conclusion was reached that the lunar surface, at a 7-to-11 inch depth, was similar to the properties of medium density soil on Earth.

A radiation device also measured the density of the very top of the lunar surface finding its density to be rather low, about 1 gram per cubic centimeter—that is, about as dense as water. Other radiation characteristics were measured and found to be so low as to present no hazard to man at all. It was revealed that *Luna 9* also carried the latter radiation devices and hence the *Luna 13* data was confirmatory. Several clear, crisp panoramas of the lunar surface were returned to Earth via the *Luna 13* telemetry system showing large rock-strewn plains all the way to the horizon in all directions; not too different from the panoramas returned by *Luna 9*.

The batteries of *Luna 13* petered out on the evening of December 30 and so the year ended on a pleasant note. The jigsaw puzzle that was the Moon for so long was slowly being filled in, and man was learning nature's secrets kept from him for eons beyond time. Such secrets are not kept by virtue of any effort on the part of nature but rather represent the state of man's ability to investigate whatever area it is that stirs his imagination and curiosity. His technical capability, his insight, and his ability to proceed in a logical manner are some of the characteristics that enable him to add to the world's fund of knowledge and material improvements.

The popular term for this state is called technology. Many is the person who wrinkles his or her nose at this term but in fact these same people ranging from members of Congress—particularly some unusually vociferous and unknowledgeable senators—to the pathetic "street people," would raise all sorts of hell if they had to revert to outhouses, do without their automobiles, be deprived of their television sets or portable transistorized radios, be denied their electric typewriters, go back to small screen movies, have all of the programmed household appliances removed (washers, dryers), travel by propeller airplane over transcontinental distances, and not be able to communicate over long distances at much lower than cable rates by using the communications satellite in synchronous orbit. If technology has brought us the hydrogen bomb and the intercontinental missile it also has brought a source of electric power that is the only answer to the depletion of fossil fuels and it has brought us reconnaissance satellites, which have brought a degree of stability to the military situation among the big powers unrealizable without them.[54]

The gap between *Luna 13* and the launching of *Luna 14* on Sunday, April 7, 1968, at 1009 GMT was too long for the Soviets not to have tried launching in the 15-month interval. No satellite that reached Earth orbit in that interval had the recognizable characteristics of a lunar lander or lunar orbiter. *Zond 4*, launched on March 2, 1968, is another tale and does not form an exception to my statement. However, I am still inclined to believe that lunar missions were tried; how many remains a question, they simply did not get to Earth orbit and so received no identifying numbers from either SPADATS (the North American Air Defense Command tracking network) or the United Nations international designations for space objects. *Luna 14* was the last of its kind, for the following spacecraft were to be much larger, more sophisticated, and of considerably greater interest.

---

54 Read *Secret Sentries in Space* by Philip J. Klass, Random House, 1971.

Again, long after the fact it was revealed that, in addition to its other payload, *Luna 14* carried prototype wheels of the Lunokhod in order to test their operation in the lunar environment. They worked well we are told aboard the Moon-orbiting spacecraft. Photography and studies of the physics of the Moon were apparently the remainder of the *Luna 14* mission.

For some long time, I wondered about the operations that occurred to place the Soviet spacecraft in lunar orbit. How much thrust did the retro-engine have? How much propellant was used? How long did the engine burn (operate) to perform the maneuver? What did the spacecraft weigh in lunar orbit after the maneuver? The Soviets never explicitly gave answers to all those questions. After reviewing once again the statements on several of the lunar spacecraft published by *Pravda*, I could only whisper to myself, "Thank you, *Pravda*" for I had discovered enough information to get good answers to all of those questions. I decided that that tidbit ought to be made known to as many members of the astronautics community as I could reach so I had the results published in *Spaceflight*, a journal of the British Interplanetary Society, of which I am a Fellow. The information collection ran like this:

> From the *Moscow News* (supplement to *Pravda*, issue No. 47 (830) of Saturday, November 19, 1966, an English language publication) we are told that:
>
>> *Luna 12* used a liquid rocket engine for "braking" into lunar orbit (page 6).
>>
>> The velocity change was from 2085 meters/sec at approach to 1,148 meters/sec at insertion into orbit. The change, therefore, was 937 meters/sec (3074 fps) (page 8).

The engine worked for a preset time, 28 seconds (page 8).

From *Moscow News,* April 6, 1966, *Luna 10* weighed 3,527 lb.

From *Pravda,* August 25, 1966, *Luna 11* weighed 3,616 lb.

I averaged the two lunar spacecraft weights and got 3,571 lb. (For the technically minded person reading this, I used a specific impulse of 300 lb-sec/lb; a conservative choice for Soviet propulsion.)

There is a very basic equation used in rocketry wherein this data can be used, and so I did. The results quickly told me that 970 lb of propellant was used in slowing down the spacecraft to enter lunar orbit; the propellant was used at a rate of 34.7 lb per second; and the engine thrust was 10,400 lb—a surprisingly large number—therefore the totality of spacecraft remaining in lunar orbit was 2,600 lb. Search as I may, and have, I've never seen any other calculations for the same items. I am quite certain that mine are correct to within 2-3 percent.

So closes the chapter on early Soviet lunar landers and lunar orbiters.

7

Soviet Reconnaissance Satellites:

The Military World

The least known aspect of Soviet spacecraft flights are those engaged in the arena of reconnaissance; photography, electronic intelligence, and communications intelligence. The latter two are frequently referred to in aerospace publications (for instance, *Aviation Week & Space Technology*) by the terms ELINT and COMMINT, whose derivations are obvious. By analogy I will call photographic intelligence PHOTINT, although I've never seen it used in print anywhere, or referred to verbally, for that matter. The trade press has often speculated about the PHOTINT capabilities of the Soviet Union, attributing to them an ability to discern objects as small as 5 feet or less—a not unreasonable number if one wishes to identify military items with some accuracy. With all of the reconnaissance spacecraft that they have launched, and recognizing the huge capital investment that such satellites represent, I have no doubt that much research in the Soviet Union is dedicated to the improvement of their photographic capabilities.

Theory tells us that a resolution of objects 1½ to 2 feet long (or wide, etc.) is possible but is limited by atmospheric shimmering. This phenomena is caused by both the motion of the atmosphere (yielding winds and therefore density changes) and the temperature variations in different atmospheric layers. Together this gives rise to changes in the refractive index of the air and causes a variety of light reflections and dispersions. The total result is a defocusing effect that impairs the quality of the image seen by the satellite cameras. Advances in photography encompassing the use of lasers as in holography (a kind of three-dimensional photography) will, in time, vastly improve the quality of satellite-taken pictures. It seems reasonable to surmise that such use is already underway but is simply not discussed in the open literature.

Despite the great hush-hush nature of this activity, the orbit parameters of all Soviet reconnaissance spacecraft are readily published in the Soviet press and have been for years, as are the characteristics for all other Soviet spacecraft. Reconnaissance spacecraft are launched under the cover of the "Cosmos program" and are not specifically identified as to functions or missions.

The magazine *Aviation Week & Space Technology* has often claimed that the photoreconnaissance satellites were modified Vostok spacecraft having cameras installed in place of the cosmonaut and all of his supporting equipment. The Vostoks were 10,400 lb spacecraft in Earth orbit. The returned manned capsule was more than half of that weight, some 5,500 lb. That represents a substantial amount of weight devoted to cameras and related equipment—even if say 3,000 lb were identified as capsule structure, parachutes, ablative material for thermal protection on reentry, and the various control systems.

Recovery of the Vostoks has been adequately explained in the Soviet (and American) press. There is no reason to believe that the reconnaissance Vostoks are recovered any differently. After its mission is completed and the correct geocentric (orbital) position is attained, the spacecraft is properly oriented, the liquid-propellant rocket engine is fired for an appropriate interval, and retro-engine compartment and extraneous instrumentation are separated, and the recoverable compartment continues on its now "decaying" orbit. After about a 5,000-mile path, it enters the atmosphere, slows down, deploys a parachute, and descends to Earth. The descent is almost always in the near vicinity of Karaganda, a large and important industrial center in Siberia located at about latitude 50° north and longitude 73° east.

Occasionally vehicles are recovered in other areas; for instance, near Kostanay or near Magnitogorsk or Orenburg. (The latter is the area where Cosmonaut Vladimir Komarov met his untimely and tragic death in the crash of *Soyuz 1*.) Whether these latter cases are generally deliberate or due to some program change during the flight program I do not know. Some possibility of an occasional malfunction (say, incorrect spacecraft orientation at retrofire) exists also. Except for the first two manned spacecraft (Gagarin, Titov) all of the other normal flights have also descended near Karaganda.

When the reconnaissance program was in its initial stages the Soviets experimented with stay-time in orbit up to periods of 16 days. Moreover, all vehicles were on 65° inclination orbits. Perigees varied little from a nominal 127 sm altitude and orbital periods were within a minute or so of 89 minutes. By the time *Cosmos 28* was launched on April 4, 1964, they had settled on a standard eight-day orbital stay-time and with some few exceptions stayed. with that interval for some years until late March 1968, when on the 21st, *Cosmos 208* was launched and stayed in orbit for 12 days thus inaugurating a new stay-time. This has now been expanded to 13 days and occasionally increases to even greater intervals.

The spacecraft are launched so as to appear over the United States—or other selected areas—during daylight hours, which means a launch from Tyuratam or Plesetsk in the early or late afternoon, local time, respectively. Recovery was, for the 65°, eight-day missions, never ever more than one or two orbits different from a standard 126 orbits. Recovery is initiated for the 65° spacecraft over mid-Africa and ends up, as I have noted, in the Karaganda region.

There are two anecdotes that are worth telling about the location of the launch bases Tyuratam and Plesetsk. In the early and mid-1960s the maps of the Soviet Union published by the National Geographic Society depicted a launch base at a point just northeast of the city of Aral'sk on the northeastern corner of the Aral Sea. Such bases were depicted by the imprinting of a small red rocket on the map. In the late 1960s, the then recently retired Air Force General Bernard Schriever was made a member of the society's Board of Directors.

Immediately thereafter, all new maps of the Soviet Union depicted the red rocket at its proper location near the town of Tyuratam at latitude 45.96° north and longitude 63.52° east. When *Cosmos 112* was launched into a 72° orbit on Thursday, March 17, 1966, at 1034 GMT it created a puzzle, for the SPADATS bulletins showed a zeroth node crossing the equator at longitude 354.24° west (longitude 5.76° east).[55] This did not fit a launching from Tyuratam, the site of all reconnaissance spacecraft launchings before that date.

I carefully plotted the *Cosmos 112* orbital trace on a map and came very close to the town of Plesetsk as the launch site, when the trace, the launch time, and the zeroth node were all considered. (See Chapter 3, "When Are They Launched?") I was skeptical of my results because they led me to a site I thought would be too far north for regular use as a satellite-launching base (long winters, cold weather); Plesetsk is at latitude 62°54' north and longitude 40°39' east. I was wrong in that skepticism; physics students at the Kettering Grammar School in England under the direction of Geoffrey Perry announced, via the press, that they had "discovered" the site some days later. I was pleased at the confirmation of my findings but chagrined at my own lack of confidence in the event.

Up until the end of 1965, all reconnaissance flights were made from Tyuratam. In early 1966, the first reconnaissance flight was made from Plesetsk and thereafter the Plesetsk launch complex played an increasingly more important role in the conduct of military oriented unmanned spaceflights. The degree of military orientation in the Soviet spaceflight program is not entirely clear. Military experiments are generally passive and unprovoking in nature, as one should expect in order to not make overt breaches of the peace. Except by examination of the experiment equipment and the spacecraft from which it operates, a policy not in keeping with Soviet views, outsiders know little about such activities.

Table IV indicates the numbers of photoreconnaissance spacecraft launched from Tyuratam and Plesetsk up to the end of 1971. No flights of this kind were launched up to the end of 1961 (except that one should consider the possibility that initial photography likely

---

55 SPADATS bulletins use astronomers' "language" and measure longitude from Greenwich to the west, only.

was conducted for military purposes aboard the Vostoks and their precursor canine-carrying spacecraft).

TABLE IV: LAUNCHINGS OF PHOTORECONNAISSANCE SATELLITES

Year	Tyuratam	Plesetsk
1962	5	0
1963	7	0
1964	12	0
1965	17	0
1966	15	6
1967	8	14
1968	14	15
1969	13	19
1970	13	16
1971	13	15
1972	10	18
1973	8	27
1974	10	19
1975	12	22

PHOTINT satellites are all recovered, but other types of reconnaissance vehicles send their data back via telemetry; the latter are eventually incinerated during decay. Thus, identification is difficult for all of these other classes. A large number of spacecraft launched from Plesetsk, the Vandenberg[56] of the Soviet Union, at an inclination of 71° and with a typical perigee of 175 miles, an apogee of about 325 miles, and a period of about 92 minutes appear to be collectors of ELINT, which may or may not include

---

56 Vandenberg Air Force Base in California was the space launch base from the West Coast of the US.

some COMMINT activity. From down here on the ground, without very sophisticated radio and radar equipment, to say nothing of code interpretations, we remain without even the most miniscule confirmation of our estimates in this case.

In like fashion, a fairly large number of satellites launched from Plesetsk at 74°, and at quite circular orbits ranging from 350 to 750 sm and having periods from 95 to about 110 minutes, perform what may be navigation functions. Some of them may be military communications satellites. There have been numerous groups of 74° satellites at circular orbits of slightly over 900 miles, with periods of 115 minutes.[57] These groups of eight satellites are likely similar to some early US defense communications satellites (placed in synchronous orbit but the Soviet Union has selected another approach). There have been some 57 satellite launchings from Kapustin Yar (KY), which is a lesser launch base about 60 miles south-southeast of Volgograd. From KY, the *InterCosmos 1-5* launchings were obviously dedicated to scientific studies among a cooperating group of Eastern European nations as was *Cosmos 261* (it could be called *Intercosmos 0*).

Of all the others, a goodly number are identifiable as carrying out scientific experiments in the area of solar research, ionospheric studies especially in electron density, micrometeorite investigations, and cosmic ray studies for rays originating beyond the solar system. However, a large number of these small satellites—I estimate their weights in the 400 to 2,000 lb range—must have had other missions. The military area is most likely but again the specific purposes on a per satellite basis remain rather unknown.

None of the Kapustin Yar satellites have ever been recovered so that whatever information they collected was returned via telemetry. The

---

57 There have been 13 such groups launched through the end of 1975.

larger ones (whichever they were) could have been used to examine cloud cover as predecessors for reconnaissance as indicated by Philip Klass in his book.[58] But any correlation with reconnaissance launches gives results for such conclusions that are far less than desirable—in short, the results are nonsensical in view of the fact that cloud motion is so frequent such launches would have to precede reconnaissance by hours not days as has been the case.

Now if these satellites collected cloud cover information over several years then a good statistical basis could be formed for the launching of PHOTINT satellites to examine a particular area whether it is in the United States, or observing French nuclear tests in the Pacific, or Chinese nuclear tests at Lop Nur in Xinjiang plus other Chinese activity at its embryonic launch bases, or the massing of equipment and troops at borders in international disputes, or the preparation of launchings of all kinds at the bases of other spacefaring countries.[59]

Klass rightly makes an excellent case for the PHOTINT examination of the missile bases, silos for missiles (whether occupied or not), and other military installations of other countries as a contribution to the stabilization of international affairs. When each of the large countries are fully aware of the war-making capabilities of the others then the probability of aggression in ignorance becomes perhaps, vanishingly small; a pleasant thought. Some so-called "famous" missile gaps of the past tended to be very destabilizing until the advent of the PHOTINT satellites.

When the Plesetsk base initiated activity, it introduced a whole new spectrum of launching angles for reconnaissance satellites. In addition to the 65° and 52° launchings from Tyuratam (*Cosmos 32* was the first 52° bird) there were now launchings at angles of 72.9°,

---

58 *Secret Sentries in Space* by Philip J. Klass, Random House, 1971.

59 The launch complex for the People's Republic of China is at Shuang Cheng-Tsu, located at 100°11' east long., 41°5' north lat.

81.2°, 65.4°, and 62.8° from Plesetsk. Because the reconnaissance launchings have become so standardized just a fraction of a degree difference in the orbit inclination angle is frequently the first indication that some other mission is underway. That is why the launching of *Cosmos 243* created a stir among knowledgeable persons.

Cosmos 243 came from Tyuratam but had an inclination of 71.3°, never previously seen in that base's activity. It stayed in orbit for 11 days during its reconnaissance mission and ejected a capsule with scientific equipment for ice-cover studies. Using both IR detectors and centimeter band receivers, detailed examination from orbit for the first time was made of ice coverage boundaries in the Antarctic and other areas of this planet. Several scientific papers were published by the Soviets on the results as well as the conduct of this series of experiments. The carrying of extra "capsules" for additional experiments was initiated on reconnaissance satellites commencing with *Cosmos 208*. Both the initiation of the long period reconnaissance spacecraft and the appearance of the added capsule occurred simultaneously. This indicated that the Vostok booster, used for such launchings, had been uprated.

Because the scientific experiments are not likely to be permitted to compromise the PHOTINT mission, the booster payload capability obviously was increased. The Vostok booster, displayed at the Le Bourget Aerodrome in France in 1967, is the original reconnaissance satellite booster. Further, the PHOTINT payload has been described as contained in a modified Vostok recoverable module. The Soviets have described the Soyuz as weighing 14,000 lb (later models are heavier) hence the Soyuz booster derived from the Vostok is required. Readily available photographs reveal that the Soyuz booster upper stage, radically changed from the Vostok stage, carries much of this uprated capability. Additional capability is to be found in the clustered first stage; both in the conical strap-on units and in the central core, the sustainer. The four strap-on rockets and their core are known as parallel staging and the core and its upper stage are termed tandem staging.

The Soviet PHOTINT reconnaissance program, up to the end of 1975, included 328 recoverable satellites; 157 from Tyuratam and 171 from Plesetsk. To the end of 1975, there have been three evident failures; *Cosmos 50*, *199*, and *758*. Toward the end of its eight-day flight, *Cosmos 50* separated into more than 12 pieces. Ground control destroyed it because a presumed malfunction prevented proper operation of its recovery system. Such action assured that the payload would not fall into "unfriendly hands" after a natural decay. *Cosmos 199*, an intended extended-time-on-orbit satellite failed as *Cosmos 50* did and was destroyed on orbit 152. *Cosmos 758* belonged to an exclusive group of satellites using a 67° inclination. Launched on September 5, 1975, it exploded the next day. Its two companions using that inclination, *Cosmos 805* and *844*, were launched in 1976. The former was recovered after 20 days and *Cosmos 844* was destroyed in orbit three days after its launching. Clearly those satellites carried a high priority but not identified payload.

The Fractional Orbit Bombardment Satellite

On September 17, 1966, a satellite was placed in orbit by the Soviet Union, which experienced a spectacular malfunction. The vehicle, contrary to the usual procedures, was not identified in the Soviet press in any manner—it was given the international designation 1966-88A and subsequently 1966-88B, et al. (yielding some 52 objects) because of the numerous fragments that resulted from the explosion of the satellite. The orbital parameters were also unusual, since perigee was only 85 miles, the lowest ever for a Soviet satellite. Apogee was somewhat indeterminate since the force of the explosion had scattered the fragments over a wide altitude range (this could have accounted for the low perigee too but didn't, as we shall see). There were pieces all the way up to 537 miles. An additional new feature was the inclination of the orbit at 49.6°. A hasty conclusion would have correlated this new angle with an almost direct easterly launch from Kapustin Yar. But when I examined the zeroth node (the prime equatorial crossing of the orbit), it became immediately evident that KY was not the origin of this unnamed and very mysterious satellite, indeed.

Again, on November 2, 1966, (1966-101A) another satellite with similar circumstances appeared and, again, all of the odd characteristics of the earlier satellite, including the explosion, were present. The whole business reminded me of the 1962 September, October, and November launchings of probes to Venus and Mars wherein five of six spacecraft failed—exploded in Earth orbit and remained unidentified by the Soviet Union. There was a Mars launch window in early 1967 but these launches were far too early to fit into that picture. No amount of guessing brought forth any fruitful results that focused on the purposes of these two missions.

On January 25, 1967, at 1355 GMT, a satellite was launched from Tyuratam at an angle of 49.6°; it had an apogee of 130.5 miles and a perigee of 89.5 miles. Its parameters were published in the normal manner by TASS, the Soviet news agency, and they revealed one more oddity about these satellites; no orbital period was given. The satellite was called *Cosmos 139*. The lack of an orbital period was of particular interest and a clue to the purpose of the spacecraft.

The evidence indicated that although it was actually placed in orbit, the satellite did not completely circumnavigate the Earth, but was deorbited toward the end of its first orbit.

Total time from liftoff to landing appeared to be close to 100 minutes assuming normal reentry procedures. The assumption proved to be wrong. By the time several more such birds were launched, enough news had found its way into the press of the Western world to excite military and political leaders in several countries. Not all of the launches that followed *Cosmos 139* were successful; Table V gives a list of all such launchings and my indication of their success. Success being "retrieval" (deorbit) after less than one orbit.

TABLE V: FOB LAUNCHINGS

Vehicle	Launch Time (GMT)		Apogee (sm)[60]	Perigee (sm)	Inclination Degree	S/F
1966-88A	Sept 17, 1966	(?)	537 (?)	85	49.6	F
1966-101A	Nov 2	(?)	344 (?)	87	49.6	F
Cosmos 139	Jan 25, 1967	1355	130.5	89.5	49.6	S
Cosmos 160	May 17	1608	127	88	49.6	F
Cosmos 169	July 17	1647	129	89.5	50	S
Cosmos 170	July 31	1647	129	90	50	F
Cosmos 171	Aug 8	1609	137	90	50	S
Cosmos 178	Sept 19	1447	127	90	50	S
Cosmos 179	Sept 22	1407	129	90	50	S
Cosmos 183	Oct 18	1331	132	90	50	S
Cosmos 187	Oct 28	1314	131	90	50	S
Cosmos 218	Apr 25, 1968	0048	131	89.5	50	S

---

60 Statute miles.

Vehicle	Launch Time (GMT)		Apogee (sm)[60]	Perigee (sm)	Inclination Degree	S/F
Cosmos 244	Oct 2	1337	132	87	50	S
Cosmos 298	Sept 15, 1969	1303	132	87	50	S
Cosmos 316	Dec 23	0927	103.5	101	49.5	?[61]
Cosmos 354	July 28, 1970	2204	129	89.5	50	S
Cosmos 365	Sept 25	1410	131	89.5	49.5	S
Cosmos 433	Aug 8, 1971	2347	161	99	49.6	S

The Soviets, over the early calendar period of these launchings (after *Cosmos 139*), now announced that they had an unstoppable rocket weapon that could attack targets from any direction and with little likelihood of detection before reaching such targets. This sort of news was the kind to make the military ranks uneasy for no ballistic missile early warning system (BMEWS) or directed energy weapons (DEW) lines of radar guarded the US, for instance, from the south.

The remarks about little likelihood of detection had interesting implications, for that meant descent to an intended target in a short interval. But how short? I had not long to wait for an answer. The then US Secretary of Defense Robert McNamara was quoted in a *New York Times* article as stating that the FOB descended from its (100 mile) altitude in 3 minutes. That was a very informative statement.[62] I questioned some of my colleagues, who were deeply immersed in studies of orbital mechanics, on the required retro

---

61 There is no way that I can determine if something new was being tried or an overburn into orbit occurred (unlikely) or the payload propulsion system malfunctioned, burned partially, and placed the satellite in that orbit. Without more data one could only guess.

62 No reference given in original document for this quoted text.

velocity necessary to bring down a spacecraft in 3 minutes from that close-in Earth orbit.[63]

From previously plotted curves, the first answer was 5,000 fps and shortly came an improved answer of 5,200 fps, a very substantial velocity increment. It was reasonably clear now that the first two attempts ending in explosions had suffered from explosions in the retro-system, where the amount of propellant for a delta V of 5,200 fps could, and did, cause one hell of a bang! Twice! Moreover, it followed that after such velocity removal, the range along the Earth's surface was only 500 miles. Compare those numbers with the values set for returning a manned spacecraft from orbit; a retro velocity of about 500 fps and an Earth range of 5,000 miles together with a time from retro to landing of about 25 minutes, in order to get a "feeling" for the time available to detect, track, and destroy the FOB. If one feels that 3 minutes is a damn short time in which to do anything, then my message has come through.

The FOB on the other hand is faced with a severe control and guidance problem during its atmospheric reentry procedure; there is little time for path correction. As a consequence, I conclude that it is almost certainly intended as a destroyer of "soft" targets; an area weapon as opposed to a point target weapon. Cities are soft targets. The soft target premise implies a warhead capability of one to five megatons. Such a warhead plus the 5,200 fps retro velocity requirement places the FOB payload-warhead, plus retro-propulsion, in the spectrum of 10,000 to 12,000 lb in near Earth orbit.

After the initial rash of FOB launchings in 1967, the Soviets settled down to a much slower pace of launchings as indicated in Table V. An FOB launching still manages to make sufficient "noise" to warrant

---

63 I revised my liftoff to landing time to 87-90 minutes.

a newspaper article but the original widespread public debate has completely vanished. The fickle public has a short memory.

The Orbit Changing Satellites: Prelude to the Orbital Destroyer Spacecraft

Almost from the beginning of the Space Age, many commentaries have appeared in the press noting the multitude of possibilities for space warfare. Many such discussions tended to be (and still are) written with a cavalier disregard for orbital mechanics and were overly romanticized, depicting opposing space forces engaged in battle as extensions of aerial combat in the manner of the Battle of Britian in 1940.

The public is still largely unaware of the economics of making large maneuvers from orbit to orbit; altitude changing is relatively simple and inexpensive while making changes in orbit plane inclination angle is complex and expensive. With the present chemical propulsion systems this situation is likely to remain unchanged.

Imagine the surprise then, of even the astronautics community, when on November 1, 1963, the Soviet Union announced the launching of a spacecraft they called *Polyot 1* (*Flight 1*) and informed the public that *Polyot 1* had performed several maneuvers in orbit, including considerable orbit plane changes. *Polyot 1* was launched at 0856 GMT and placed in an 891 by 213 sm orbit at an inclination of 58.5°. It was the last number that gave rise to much controversy.

The first SPADATS bulletins were inaccurate in displaying the numbers describing the orbit; later bulletins gave rather different values for the inclination angle. The nub of the discussion finally dwindled down to whether or not plane changes were made after *Polyot* was in orbit or had the changes been made during boost to orbit? The latter are far less costly. The issue remains unresolved, at least in the public domain.

Polyot 1 was followed by *Polyot 2* on April 12, 1964, launched at 0931 GMT into a 310 by 192 sm orbit at an inclination of 58°. Since this day was the third anniversary of Gagarin's historic spaceflight, I thought that a Niagara of information would be forthcoming on the Polyots. 'Twas not to be. Outside of remarks that were similar to those connected with *Polyot 1*, nothing.

Cosmos 185 was launched on October 27, 1967, into what the Soviets announced as a 551 by 324 sm orbit at an inclination of 64.1°. The bulletins that followed its launch described the orbit just as announced by the Soviet Union. What I didn't learn until later on (in the *New York Times* of Wednesday, April 3, 1968) was that *Cosmos 185* had made a substantial altitude maneuver shortly after its initial orbit was achieved. The change it made could be guesstimated from Table VI, enumerating changes made by other maneuverable satellites.

Table VI: Altitude Changing Satellites

Cosmos	Launch Date Time, GMT	Initial Orbit, sm	Later Orbit, sm	Inclination Degree	Period, Minutes	
					1st	2nd
185	Oct. 27, 1967 0214	175 x 160[64]	551 x 324	64.1	89.7[65]	98.7
198	Dec. 27, 1967 1127	175 x 165	590 x 556	65	89.8	103.5
209	Mar. 22, 1968 0929	175 x 155	587 x 541	65	89.6	103.1
217	Apr. 24, 1968 1600	164 x 88	323 x 246	62.2	88.54	93.4
291	Aug. 6, 1969 0540	357 x 95	—[66]	62.3	91.5	

The velocity increment required for this orbit change for each satellite was approximately 550 fps, which could be interpreted as 500 to 600 lb of propellant for a 10,000 lb spacecraft. Not really too much for so respectable a maneuver.

The next time satellites of the Soviet Union would maneuver on orbit John Q. Public was going to be impressed, indeed.

---

64 Estimate.
65 Estimate.
66 This test evidently didn't work completely. It appears that one velocity increment was added but that the second failed. Bulletins did not indicate an orbit at any time different from the one shown although there ought to have been one with a lower apogee.

The Maneuverable Satellites: A Second Look

The Soviet Union evidently had more than one use for the class of satellites I have termed "maneuverable." Slipped in between the events of the anti-Cosmos defense force activity, *Cosmos 367* was launched on October 3, 1974, at 0959 GMT. Little time passed before it was recognized as having characteristics similar to those displayed in Table VI. Initially, it was at 65° and in a 165 by 150 mile orbit; sometime shortly afterward, it climbed to a 640 by 579 mile orbit still retaining its original inclination.

As a consequence of this launching, the maneuverables clearly called for another look, for this bird was not of the anti-Cosmos defense force and, moreover, had come more than a year after its immediate predecessor. I now reexamined all prior launchings to see if, perhaps, some other satellites, previously unrecognized, belonged to this somewhat mysterious grouping. While its orbit parameters were distinctly different in their "final" form, it came to my attention that *Cosmos 125* had maneuvered into a circular orbit of 155 miles from a somewhat lower apogee/perigee—it did not maneuver again noticeably. It appeared that this was the first cautious attempt in the maneuvering area—separate from the Polyots. It is possible, of course, that the orbit of 155 miles resulted from a failure in the propulsion system—that remains an unresolved question. Moreover, it is not unreasonable, in view of the peculiarity of the orbit parameters that *Cosmos 125* belonged to a different, unrelated class of spacecraft. In that light its intended purpose remains rather unclear.

Table VI-A delineates the same type of data for *Cosmos 367* and the nine following satellites that is given in Table VI for the earlier maneuverables. While the parameters for the several spacecraft seem consistent enough, there is no obvious periodicity that can be determined from either launch times or dates. Thus, linking the mission with time of day or part of the year has not yielded any additional information.

However, by the time *Cosmos 469* and *516* were launched several suggestions had been put forth indicating a belief that these satellites were being used for the purposes of sea surveillance. One of the first to offer this explanation was Philip Klass, senior avionics editor for *Aviation Week & Space Technology*. I would add to this my own surmise that sea photoreconnaissance is the subject of interest when the spacecraft is in its initial lower orbit and that the use of side-looking radar then proceeds at the higher altitudes, later on.

Ships have identifying signatures, like any other source that is emanating energy. Thus sizes, fuels being consumed, electromagnetic radiation, the geometry of a single ship, and the juxtaposition of several ships (as in a convoy or in naval vessels engaged in war games) all contribute to such signatures. The sophistication of side-looking radar has reached the point that most of these characteristics could be resolved through its judicious use. IR optics is quite evidently a requisite adjunct to the radar.

Given that passage for most commercial ships can largely be ascertained by appropriate reading of the newspapers, they can be relegated to a secondary, though not dismissible, place in the list of desired observables. Naval ships are obvious targets. In particular, investigations have been conducted on surface effects due to submerged submarine movement. Indications are that these effects can be recognized from orbital altitudes. Also, given the importance of petroleum the large oil tankers provide targets of major interest.

Table VI-A: Altitude Changing Satellites

Cosmos	Launch Date Time, GMT	Initial Orbit, sm	Later Orbit, sm	Inclination Degree	Period, Minutes 1st	Period, Minutes 2nd
125	Jul. 20, 1966 0858	?	155 x 155	65	?	89.5
367	Oct. 3, 1970 0959	165 x 150	640 x 579	65	89.5	104.5
402	Apr. 1, 1971 1128	170 x 154	642 x 590	65	89.72	104.95
469	Dec. 25, 1971 1128	172 x 161	634 x 584	65	89.7	104.7
516	Aug. 21, 1972 1034	172 x 159	640 x 572	65	89.6	104.3
626	Dec. 27, 1973 2018	174 x 160	615 x 565	65	89.6	104.4
651	May 15, 1974 738	172 x 159	593 x 554	65	89.5	103.4
654	May 17, 1974 0651	172 x 162	636 x 567	65	89.7	104.4
723	Apr. 2, 1975 1058	172 x 159	591 x 569	65	89.6	103.74

Cosmos	Launch Date Time, GMT	Initial Orbit, sm	Later Orbit, sm	Inclination Degree	Period, Minutes	
					1st	2nd
724	Apr. 7, 1975 1058	172 x 160	582 x 540	65	89.6	103.04
785	Dec. 12, 1975 1243	161 x 155	634 x 557	65	89.6	104.3

THE SOVIET ANTI-COSMOS DEFENSE FORCE

When *Sputnik 1* was launched and overflew every country on this planet, a precedent was established for all future satellite flights. Nevertheless, the proclivity of the major space powers to indulge in military activity that they prefer to keep under wraps has prompted the development of protective measures to assure themselves of the complete integrity of their operations on orbit. Many such measures must be passive; for instance, some kind of armor around sensitive instruments to protect against small projectiles, thermal protection against heat rays (lasers), and design against nuclear radiation either from manmade (nuclear weapons) or natural sources.

On the other hand, suppose that some country decided to not only perform orbital snooping by examining close-up another's military spacecraft but also to sabotage it in some manner. In that case the would-be "victim" ought to have been designed to protect itself. This would require a means for detecting the presence of the offending spacecraft as well as a method for determining that it is indeed hostile in intent. It should be remembered that American and Soviet naval units have played "chicken" on the high seas and, despite some bruising incidents, no hostile actions have ever been taken by either side. Consequently, it is fundamental that before a commitment to hostile action is made, one must be sure that it is not just a game of orbital chicken that is being enacted. Otherwise, the resulting orbital chaos could cascade from the game of chicken to the reality of a space war.

All of this being so, the Soviet Union launched *Cosmos 248* from Tyuratam into a very circular orbit at 0441 GMT on October 19, 1968. This spacecraft in a 94.8 minute orbit, 342 by 304 sm, had a new wrinkle; its inclination was 62.3°, seen before only on the occasion of *Cosmos 217*.[67] The new inclination indicated that some new program or test was in process. So, it was.

The killer satellite whirled in its orbit waiting for the next day when *Cosmos 249* would be launched, also from Tyuratam, at 0402 GMT, a time selected to place both satellites in essentially the same orbit plane. The usual TASS communique depicted *Cosmos 249* in an orbit of 1338 by 306 sm at an inclination of 62.4° but in fact the satellite had initially been placed in a very low orbit of about 80 sm altitude, boosted to 312 by 1017 sm orbit and then placed in the announced orbit.

During the course of its maneuvers *Cosmos 249* passed very close to *Cosmos 248* and suddenly there were several dozen pieces of debris in the path previously occupied by that satellite. The Soviets' long,

---

67 It is possible that *Cosmos 217* was a failed "killer" satellite of the anti-Cosmos defense force.

drawn-out tests with their maneuvering satellites had led to this first successful interception. There was more to come.

At 2351 GMT on October 31, 1968, *Cosmos 252* rose from its pad at Tyuratam and soon was in orbit. Its original orbit I have not been able to ascertain, but the TASS communique in the November 2, 1968, issue of *Pravda* gave the orbit parameters as 1350 by 334 sm at 61.9°. In addition to the standard release the TASS statement also had one additional line which stated, "The scientific investigations provided by the program have been performed."

That was a clue that even the sophisticated writers of the *New York Times* did not pick up; for writing in the *Times* of February 8, 1970, Richard Lyons referring to these tests claimed that, "Sometime within the next 13 days [after the *Cosmos 252* launch]...the perigees of 248 and 252 crossed and the latter blew up." Since TASS stated that "investigations...have been performed" on November 2, 1968, it follows that the intercept/destruction must have occurred by that time. Indeed, they did!

As *Cosmos 248* commenced its 198th orbit it crossed the equator at longitude 18.570 west at 0342.63 GMT on the morning of November 1. Keeping it company close by was *Cosmos 252*, starting the beginning of its third orbit; it crossed at longitude 18.31° west at 0342.81 GMT, both were heading toward radar sight of the Plesetsk launch complex. The deadly game of orbital tag was in progress and at the appropriate signal from Plesetsk, there were many pieces where *Cosmos 252* moments before had whirled above our planet. In the *Satellite Situation Report* of October 31, 1971, issued by the Goddard Space Flight Center, there were listed 76 "objects" associated with *Cosmos 249* and 89 "objects" associated with the *Cosmos 252* launching. Interpret objects as debris or remains, if you will; the intercept and destruction of two satellites by another. The analogy to hunter-killer submarines is a good one.

I decided that the next time that this sort of test was made, I was going to follow it with increased care in order to try and find the location of the actual destruction event itself. I have no positive way of proving that I did indeed find it, but I'll let the reader decide.

The orbital acrobatics indulged in by *Cosmos 373, 374,* and *375* were considerably more elaborate than the antics of their predecessors. Each of the three satellites actually had three different orbits as depicted in the SPADATS bulletins; while the Soviets announced, in each case, the orbital parameters as those of the "final" orbit. As is usually the case, TASS announced values that differ slightly from the parameters given in the SPADATS bulletins.

Contrary to remarks in the press, *Cosmos 375* did change its orbit twice. The three sets of orbits are shown in Table VII. In the case of *Cosmos 373* and *374*: at 0751.3 GMT on the morning of Friday, October 23, *Cosmos 373* crossed longitude 19.85° west, at the equator, at the beginning of its 43rd orbit. Closely on its heels, on its second orbit came *Cosmos 374*, crossing longitude 19.91° west at 0751.55 GMT. Both satellites headed northeast and reached almost identical altitudes just north of latitude 51° north. *Cosmos 373* had a shorter period at this time so that it was in a catch-up mode (see SPADATS Bulletin 3 values in Table VII). The respective orbital inclinations were barely 0.01° apart. Tracing the two trajectories on a map the two spacecraft came into radar sight of Plesetsk and the destruction of *Cosmos 374* probably occurred shortly thereafter.

For the case of *Cosmos 373* and *375*: at 0543.55 GMT on Friday, October 30, *Cosmos 375* crossed longitude 18.61° west at the equator at the beginning of its second orbit. And once more *Cosmos 373* followed closely, crossing longitude 18.57 west at 0542.7 GMT. These satellites were also heading northeast toward Plesetsk, as had the earlier pair of spacecraft. *Cosmos 373* was then in its 148th orbit. The longitude crossings for the second pair were close to those for the first encounter as might well be expected since Plesetsk appears to be the control center and the target satellites were each destroyed during their second passes near Plesetsk, respectively.

Though the tests appear to have a primary intent of target satellite destruction it is also true that the maneuverability could play a part in rescue mission practice with quick response a critical factor.

While the exact mode of destruction is as yet unknown to me there are at least three methods that can be considered likely:

- The obvious choice of an IR or radio-guided satellite-to-satellite missile.

- Detection by the chaser (*Cosmos 373*) of the target(s) (*Cosmos 374, 375*) and a commensurate signal to detonate, onboard the target, preset explosives.

- Only slightly more exotic, the use of a laser beam to penetrate the tanks and detonate the propellant (fuel) remaining in the last stage of the booster if still attached—or in the tanks of the spacecraft itself. The Soviets have described their fuels as "hydrazine derivatives"; monomethylhydrazine (MMH), a monopropellant is a possible candidate for this operation.

Table VII: Orbital Values for the Cosmos 373, 374, and 375 Spacecraft

SPADATS Bulletin		Cosmos 373	Cosmos 374	Cosmos 375
1	p[68]	101 min	100.7 min	100.43 min
	i	62.889°	62.96°	62.856°
	hp	316.7	329	351
	ha	684.6	654	617
2	p	93.06 min	112.25 min	112.13 min
	i	62.95°	62.938°	62.839°
	hp	250.8	328.4	288.7
	ha	345.6	1323.4	1356.9
3	p	94.79 min	112.26 min	111.85 min
	i	62.93°	62.933°	62.806°
	hp	293.6	330.8	326.47
	ha	333.3	1321.7	1302.37

The TASS Announced Parameters

	Cosmos 373	Cosmos 374	Cosmos 375
p	94.8 min	112.3 min	112.4 min
i	62.9°	63	63°
hp	304.5	333	334.3
ha	343.6	1337.8	1344.64

Besides *Cosmos 248*, *249*, and *252* in 1968 and the troika of *Cosmos 373*, *374*, and *375* two other satellites in the 62° orbit have been *Cosmos 217* (April 1968) and *Cosmos 291* (August 1969). They too were maneuverable spacecraft.

---

68 p = orbital period, i = orbital inclination, hp = perigee, ha = apogee. Perigee and apogee are expressed in statute miles.

Chapter 7: Soviet Reconnaissance Satellites

The Soviet Union quite evidently intends to keep its anti-Cosmos defense force in tiptop shape for any future contingencies for there have been three more multi-satellite tests of this kind, all in 1971 involving *Cosmos 394* and *397* in February, *Cosmos 400* and *404* in March and early April, and *Cosmos 459* and *462* in late November and early December.

Several Midwest newspapers on Tuesday, December 28, 1971, carried a story about the decay of a Soviet satellite that, according to North American Air Defense Command (NORAD) alerts, was supposed to reenter in the vicinity of the town of Bad Axe, Michigan. Both Michigan state police and the Bad Axe police were notified of the expected event; the coordinates relayed gave reentry within a 200-mile radius of Bad Axe. Information was so garbled that while one location had the action within the Bad Axe region, another had decay about 100 miles east of Fairbanks, Alaska.

The actual decay on December 27 (Monday) took place in the far south Indian Ocean at latitude 65° south, longitude 144° east, near Antarctica. While the *Detroit Free Press* quoted someone as saying that "it's just a piece of junk" and the *Ann Arbor News* quoted another source saying that, "it's some scientific satellite" the fact is the satellite was *Cosmos 459*, a Soviet satellite killer. It's nice to know that coordinates for an ICBM attack were not being relayed.

8

Soyuz

The launching and recovery after 22 days of two dogs by the Soviets brought a great bubble of anticipation to those of us who had an overwhelming interest in manned spaceflight. It certainly was not clear that the Voskhod program had ended, but it was now a year since the last manned launch—Leonov and Belyayev in *Voskhod 2*. Clearly it was time for new manned space activity. But where the hell was it? More time passed without any clue in this arena until *Cosmos 133* was launched on Monday, November 28, 1966, at 1100 GMT. Its orbit was 144 by 112.5 sm, at an angle of about 52°, and with a period of 88.4 minutes, all the marks of a precursor to manned flight.

Evert Clark writing in the *New York Times* of December 8, 1966, expressed some ideas regarding this flight and a few others. Confusion reigned supreme. He had Tyuratam confused with Kapustin Yar, missed the size of the FOB and confused it with manned launches, got the wrong date for the launch of *Voskhod 2*, and interpreted the *Proton* physics experiment satellite as a "forerunner of a multi-man space bus." He did, however, note the characteristics of *Cosmos 133* as belonging to the class of manned vehicles. Finally, he, as did many others, misinterpreted the first two FOB failures as Mars probe attempts. There was a Mars launch window in late

1966, early 1967. *Cosmos 133* was in orbit for a very short period; it was recovered after 32 orbits, at 1020 GMT near Karaganda, having flown for just over 47 hours. It performed some maneuvers in orbit that were fairly substantial in terms of the delta V (delta velocity) used—about 700 fps.

Speculation was rampant—many thought a manned flight would occur within a few days and I wondered how soon Soviet cosmonauts would be in space again. Since *Cosmos 47* had been a precursor for *Voskhod 1*, and *Cosmos 57*[69] was a precursor for *Voskhod 2*, it seemed evident enough that here was the advanced calling card for *Voskhod 3*.

Because so long a time had now passed since the last flight—20 months—I expected a rather sophisticated mission and one that was bound to last a week or more. It was clearly evident to me that a multi-manned spaceflight would occur. Baloney! None of these things happened. Instead, *Cosmos 140* was launched on February 7, 1967, at the early hour of 0320 GMT, again from Tyuratam, and into an orbit not much different from that of *Cosmos 133*—the perigee was slightly lower but apogee was a bit higher; angle of inclination and period were similar. It too was recovered after some 47 hours near Karaganda after 32 orbits of flight. Besides being a precursor, *Cosmos 140* was used to perform a unique experiment in electromagnetics and superconductivity, as I noted in Chapter 13, "The Future" on scientific satellites.

I now reflected on the fact that Soviet manned spaceflight was initiated with the Vostoks, which had five prior test spacecraft before the first flight with Yuri Gagarin. Since two presumed precursors had now gone up it seemed that perhaps a new spacecraft was being tested—else why two precursors? There was nothing to do but wait and watch.

---

69 *Cosmos 57* had a major malfunction in orbit but evidently not of the kind to affect the flight of *Voskhod 2* since the flight occurred within a month after *Cosmos 57*.

A month later the Soviets managed to thoroughly baffle everyone who would render an opinion on the topic when they launched *Cosmos 146* on March 10, 1967, at 1300 GMT. This craft was placed in a 193 by 118 sm orbit, at 51.5°, and with a period of 89.2 minutes. Now this was close enough to those values that define a flight for manned spacecraft so that it initially appeared to be still another such test. The *London Evening Standard* reported that this was the largest space vehicle ever launched. American sources discounted this "analysis." Vice-President [Hubert] Humphrey, at the time, suggested that the Soviets were soon to place a large platform in space—with men—(i.e., a space station) and this remark was connected with *Cosmos 146* as a test for this space station by some writers. (The latter remarks were furthered by statements from Yuri Gagarin some weeks later, on April 8, that hinted of a manned space station to be orbited.)

The confusion coefficient was raised still higher when a few days after being placed in orbit a secondary payload was separated from *Cosmos 146*. I had thoughts of my own about this and now wondered about the possibility of a very late launch to Mars since, although the launch window was optimum during January, a fair chunk of added velocity would make up for the very late launch date in March. Sometimes it's nice to play guessing games, but it is frustrating not to ever know if the truth was approached. I figured I was stretching it here. *Cosmos 146* payloads decayed over areas where they would not be recovered: one payload appears to have decayed over the mid-Pacific while the other decayed over Siberia-China.

These notices from NORAD added to the mess; the main body was not accounted for. Later on, I learned that it had deorbited after only 17 orbits following some extensive use of its rocket engine. It also came to light that *Cosmos 146* had done some maneuvering immediately after launching as well. As it occurs, the deorbiting after 17 orbits did place the spacecraft in an excellent position to be recovered in the Karaganda area. Whether or not *Cosmos 146* was recovered remains a mystery to me, even after all these years. We'll leave it to the Russians to reveal the facts when it pleases them to do so—if ever.

Wrapped in their cloak of secrecy, the Soviets took the next step in (presumably) their manned program and in confusing Soviet watchers like me. On Saturday April 8, 1967, at 0900 GMT, *Cosmos 154* was launched into an orbit whose difference from the orbit for *Cosmos 140* was minimal. Again, as in *Cosmos 146*, there were additional pieces in orbit and again, like *Cosmos 140*, there was a recovery after 32 orbits in the Karaganda area. Rumors of a "very large payload" were repeated in the press.[70]

In the aerospace community, there are a large number of so-called marketing sheets [or newsletters] that may be publicly subscribed for and which carry all sorts of interesting information concerning business and other activities in that community. Most of them are collectors of information that are highly focused in a particular direction. Consequently, their information is usually quite good and saves one the need to ferret such material out of widely scattered sources. Some of the sheets add various rumors to their assemblage of facts and still others like to speculate on Soviet space activity.

One such marketing sheet had a writer who, from all the evidence, not only speculated far wide of the mark on Soviet activity but also neglected to pay attention to well-qualified sources on the topic like *Aviation Week & Space Technology*. As a result, he (or she) completely confused the Soviet strictly military FOB program (see Chapter 7, "Soviet Reconnaissance Satellites" on military satellites) with the tests for what was to be the Soyuz spacecraft. Those of us who could recognize the fact were amazed at the total befuddlement of this writer; more especially so in view of all of the information that was so readily available.

---

70 An analysis of the *Cosmos 146* and *154* flights is to be found in Chapter 10, "Zond Spacecraft to the Moon."

San Francisco certainly is one of the most charming cities anywhere on this planet. So it was that our neighbors, Richard and Barbara Hollis, and my wife and I decided to spend the weekend of April 22-23 in the "city." The weekend was pleasant, the weather on Sunday was not, but I found it brightened considerably when the morning news reports noted, with some excitement, the launching of a new Soviet spacecraft called *Soyuz* (it means "union" in Russian). In command was a cosmonaut on his second flight, Vladimir Komarov. He was alone but rumors flew thick and fast on an impending second launch. The rumors were well planted for they emanated, no less, from Moscow itself. Good! That gave them some substance.

Komarov was launched at 0035 GMT into an orbit of 139 by 125 sm, at an inclination of 51°40', and an 88.6 minute period. These orbit parameters fitted right into the group noted for the four birds discussed above: so much, say I, for precursors. Now for the real thing. From extensive bulletins and a thorough report in *Pravda* for the first few orbits, the news shrunk to brief, almost laconic statements. I had an uncomfortable feeling but shrugged it off as being the secretness of the Soviets; the mission after all was not complete and it was their habit when, after success, that they were informative. Perhaps it was the tendency of local radio to report local trivia.

That all was well initially is supported by the following bulletin reprinted here in part:

SOIUZ 1 EXPLORES SPACE
TASS REPORT: Flight of the New Soviet Spacecraft

Pravda, No. 114 (17796), 24 April 1967

0335

Today, 23 April 1967, at 0335:00 Moscow time, a new Soviet spacecraft "Soiuz-1" was injected into an earth satellite orbit in the Soviet Union by a powerful booster. The "Soiuz-1" is piloted by a citizen of the Soviet Union, USSR Pilot-Cosmonaut Vladimir Mikhailovich Komarov, Engineering-Colonel and Hero of the Soviet Union, who had previously performed a space flight in the "Voskhod."

The purposes of the space flight are:

> Testing of a new piloted spacecraft;
>
> Check-out of systems and elements of the ship's construction under space flight conditions;
>
> Conducting of extensive scientific and physico-technical experiments and investigations under space flight conditions;
>
> Further continuation of medico-biological investigations and the study of the influence of space flight factors on the human organism.

The "Soiuz-1" was injected into an orbit close to the nominal. According to preliminary data, the period of ship's rotation around the Earth is 88.6 min; minimum distance from earth (in perigee) is 201 km; maximum distance from Earth (in apogee) is 224 km, and the orbital inclination is 51°40'.

Reliable two-way communications have been established with the "Soiuz 1."

According to the report of the ship's commander, Comrade Vladimir Mikhailovich Komarov, as well as according to telemetry information, he went through the injection into orbit and the transition to weightlessness completely satisfactorily.

Chapter 8: Soyuz

He feels well.

Vladimir Mikhailovich Komarov has proceeded to execute the designated flight program.

Communications from the "Soiuz 1" are being transmitted on 15.008, 18.035 and 20.008 Mc.

The on-board systems of the spacecraft function normally.

Further messages on the course of the flight will be transmitted to all radio stations in the Soviet Union.

0800

The Soviet spacecraft "Soiuz-1" continues orbital flight. At 0800:00 Moscow time, the "Soiuz-1" has completed three orbits around the earth. The ship's commander Pilot-Cosmonaut Vladimir Mikhailovich Komarov performs the designated research program.

Flying over the Soviet Union, Comrade Komarov transmitted greetings to the Soviet people from on board: "On the eve of a glorious historical event, the 50th anniversary of the Great October Socialist Revolution I transmit warm greetings to the peoples of our nation, lighting the path of humanity to communism."

According to information from Komarov and from telemetry, the cosmonaut feels good, his pulse rate is 82 beats per minute, and his respiration rate is 20 per minute.

* * *

Any difficulties that manifested themselves early in the mission must have been looked at as minor and something that would be cleared up after a short time. This wasn't to be so. From the trade magazines *Aviation Week & Space Technology* and *Technology Week* (of early May 1967) and the *New York Times* of late April the substance of the problems of *Soyuz 1* can be observed. All of these reported that a major voice antenna did not deploy, that attitude control either did not work properly or had to be used excessively, that data from the ground was not received adequately (related to the antenna problem) and that because of these and other (undisclosed) problems Komarov had to give more than ordinarily required attention to these facets of the flight and so was forced to neglect the main intent of the flight—presumably, the upcoming rendezvous with another spacecraft.

Since it is now long after the fact, the antenna difficulty and attitude control problem are known to have stemmed from the same source. Photos of the Soyuz show an antenna affixed to the end of each of the solar panels of the spacecraft. Even crude estimates of its length, based on other dimensions observed, show that these antennas are used for voice broadcast. Moreover, the failure of a panel to unfold would cause attitude control difficulties since such control is designed to work best when the spacecraft is deployed in its proper working configuration. If the panel had failed to deploy, not only would there be a communication problem, per se, but also there must have been an electrical power problem as well. Komarov had his hands full. All of these circumstances led to a mission abort.

Attempts were made on the 16th, 17th, and 18th orbits to control the spacecraft attitude for retro-fire; only the latter try succeeded. Several publications tried to offer convincing stories that *Soyuz 1* tumbled throughout the retro-sequence as well as during reentry. Were that true, Komarov would have been incinerated along with his spacecraft.

It is possible that some lack of control could have occurred during retro-fire around the longitudinal axis (the roll axis) but that is all. Any other disturbance would have prevented a proper reentry; in some cases, there would be no reentry, period. The stories about tumbling during reentry are totally incorrect. Such lack of control would have caused the entire spacecraft to burn during reentry. Since the service module, the orbital module, and the solar panels are separated before reentry, it is evident that the problem that caused the parachute entanglement was related to some motion of the spacecraft after reentry was complete and was confined to the command module itself.

The parachute is deployed at an altitude of 7 kilometers (about 23,000 feet) where the spacecraft is moving at perhaps 600-700 fps (about 410-480 mph). No specific explanation of just how the chute became entangled has ever been published despite the fact that the Soviets claimed that the description of the descent in detail was given by Komarov until the moment of impact. That sort of performance, in the face of imminent death, is indicative of a coolness found in very few persons, indeed. Whatever it was he said, Komarov was trying to make certain that his accident would not befall anyone else in the future. No doubt he was also trying, up until impact, to head off his own and the Soviet space program's disaster. Alas, he failed. He gave his life in a great cause, but his death would not alter the path to the stars, only slow it a bit. <u>Yuri Gagarin had opened the door to infinity</u>; nothing would ever close it. Not the earlier end of the three Apollo astronauts[71] nor Vladimir Komarov's untimely demise.

In order to quell the myriad idiotic rumors that followed the crash of Soyuz, Colonel Gagarin, no less, gave a press interview to dispel the nonsense and firmly, officially establish the manner of the accident the parachute entanglement and Komarov's death at impact of *Soyuz 1* in the vicinity of Orenburg, Soviet Union. It must be noted

---

71 [On January 27, 1967, a fire on *Apollo 1 killed US astronauts Gus Grissom, Ed White, and Roger Chaffee.*]

that the Soviets had contributed to the rumor mill by withholding news of the accident until 11 hours after it occurred at about 0317 GMT on the morning on Tuesday, April 24, 1967; a black Tuesday.

Along with the condolences of most governments came an offer from the US to send two astronauts to the funeral. The Soviet answer was a rather cold no! They chose to be alone in their grief. A mistake, I think, but very understandable under the circumstances. The drive into space had claimed its first victim during an active mission.

Pilot-hero-cosmonaut of the Soviet Union Vladimir Mikhailovich Komarov was buried in the Kremlin Wall on April 26, 1967. His ashes, in an urn, were placed in a niche, made for the purpose and a plaque placed over the niche bearing his name birthdate and date of his demise. The planet Earth was sad that day.

The Soyuz program would see a long delay before another manned flight. Apollo wasn't the only program delayed because of the loss of crew members.

All during the time of the precursor tests and commensurate with the *Soyuz 1* flight, publications everywhere assumed that the Soyuz program was aimed at the Moon. Some of this was reinforced when the stories of *Cosmos 146* and *154* supposedly being very large spacecraft reached the public. These were presumed to be of the order of 50,000-60,000 lb. Constant parallels were drawn between the Soviet Soyuz and the US Apollo programs. Stories of Saturn V class Soviet boosters were frequent, but no evidence was ever available to show that they, indeed, had flown.

The market sheet that misinterpreted the FOB program kept saying that the Soyuz tests were practice for a circumlunar manned flight. All the trade magazines reported in a similar manner. But if the Soyuz was the command module where was the lunar booster? If a lunar landing was also in the offing, as also frequently reported, where was the lunar landing vehicle? Where, indeed, were its test flights? For the circumlunar flight where were the tests for rendezvous at Earth? Every writer in the business of reporting on space activity had a field day trying to attach one spaceflight after another to the "Soviet drive to the Moon." These stories were misplaced in time; there will be such a Soviet landing but with some other spacecraft at some other time.

There were stories of an attempt to fly the big booster, which had supposedly ended in a disaster, not in lives, only in hardware. This giant had blown up on its launching pad during a try at launching.[72] Several papers claimed that the wreckage had been photographed via satellite. The *New York Times*, on the other hand claimed, quoting Washington sources, that this was all essentially hogwash and that no such event had occurred. Who to believe? Credibility gaps were scattered all over the place.

Several years later, Kenneth Gatland, a Vice-President of the British Interplanetary Society, writing in the semi-technical journal *Spaceflight*, claimed that, indeed, American reconnaissance satellites had photographed the huge vehicle on its pad saying that it looked like a "fat bullet" and had a "large diameter"—no numbers specified.[73] Gatland also stated that the Soyuz program had as one of its main objectives the initiation of and refinement of procedures for rendezvous of a shuttle spacecraft with a large space station. The station, I assume, that was to be launched using the very large booster.

All of this was evidently going to suffer a long delay because of Komarov's fatal accident.

---

72 Or during a static test, so it was said.
73 "New Frontiers," *Spaceflight, vol. 12, 1970, pg. 166.*

Some months passed without any discernable manned activity. In August 1967, the Conference on Peaceful Uses of Space, Satellite Technology, and Law was held at Stanford University in Palo Alto. During the course of events one member of that small but supervirulent breed of anti-Soviet nuts that run loose throughout the world raised the question to the Soviet representative, Gennadi Stachevski, as to why the Soviet Union had not admitted to the world the loss of some dozen cosmonauts in addition to that of Komarov. The speaker was sarcastic, impolite, and sounded like an inquisitor in the days of witchcraft. He succeeded in embarrassing everyone in the audience including me, as I was unfortunate enough to be sitting in the adjacent seat. I felt like belting him one.

After the morning intermission for lunch, I met Stachevski for a moment and told him that relatively few persons believed the earlier nonsensical outburst. His retort was to the effect that in this day of modern communications (i.e., the combined efforts of intelligence agencies, etc.) such losses could not be hidden no matter how the Soviet Union might try to do so. I asked him later in the day if he thought enough time had passed for the manned program to resume activity. He thought that "enough time had passed...that the parachute problem had been solved." It was, in fact, to be more than another year before the manned Soyuz program would be resumed.

If the Soyuz problem(s) were to take a long time, then the Soviets had some plans to fill the gaps. Every time a satellite would go up into a 51° orbit, I would study it diligently to see if there was some relation to the Soyuz program. There were numerous 51° satellites but they were elements of other programs, mostly they were reconnaissance birds. *Venera 4* went off in June, but it was promptly announced as such and it had its own headlines to make later in the year. (See Chapter 4, "To the Planets" on interplanetary flight.)

On October 27, 1967, *Cosmos 186* was launched at 0930 GMT and it showed most of the aspects of manned flight in its orbit parameters.

Aha! I said, here is a test of the modified Soyuz spacecraft. Probably a couple of days in orbit to test attitude control, communications, solar panel unfolding, and, most important, the parachute system on recovery. The TASS announcement included remarks about the test of "new systems and structural elements of space vehicles."

Other statements gave strong indications that this was not just one more launching. After two days, however, the vehicle was still in orbit. On Saturday, the day after *Cosmos 186* went up, another FOB was launched and deorbited. Both newspapers and magazines misconstrued *Cosmos 187*, the FOB, as having something to do with *Cosmos 186* and the rapid disappearance of the FOB from orbit gave rise to tales of failure associated with the *Cosmos 186* test. They were cocksure but dead wrong. The Kettering Grammar School students, ever alert, were satisfied that *Cosmos 186* was "large enough to carry five cosmonauts." NASA officials in Washington termed the Kettering students' remarks "pure speculation." NASA claimed that *Cosmos 186* had instrumentation to study space phenomena and was no more than a scientific satellite. Subsequent events showed the Kettering students to be better informed than NASA.

Cosmos 188 was launched from Tyuratam at 0812 GMT on Monday, October 30, into an orbit almost identical to that of *Cosmos 186*. By 0920 the two spacecraft were firmly docked and were over the eastern coast of South America, over Argentina. This was the first unmanned docking in space and the first docking exercise of any kind for the Soviet Union.

When *Cosmos 188* was injected into orbit, the distance between the two spacecraft was about 15 miles and the difference in velocity was about 82 fps (about 54 miles per hour). The apparatus on both of the vehicles included restartable rocket engines to correct their orbits and make major changes in distance between them, low thrust engines for attitude control and for mooring, the docking units, and all of the appropriate radar and other electronic instrumentation for the craft to "find" each other. The main engines brought the satellites to within to about 1,000 feet of each other, then the low thrust engines took over. The latter part of the docking took place at a low velocity about 10 fps. Finally, closing and docking was slower still, at about 2-3 fps. The satellites stayed docked for 3½ hours and then undocked at 1250 GMT. The docking had taken place during the 49th orbit of *Cosmos 186* and the first orbit of *Cosmos 188*, completely out of sight of the Soviet Union, in the southern hemisphere—a remarkable achievement.

Soviet space officials were very pleased and the information flow was unusually good. Great detail on the exact methods employed in performing the rendezvous were released within a few days on November 2—and the press carried long articles on all aspects of the mission. Once again, the firm conviction that this was all preparatory for a manned Moon flight was offered by almost every writer conversant with space activity. In retrospect, I would suspect that the Soviets would have been pleased to have made as much progress toward the lunar goal as all these writers credited to them.

Models of the spacecraft were presented in *Pravda* photos but these were altered so that the true configuration (i.e., that of Soyuz) was disguised. The rendezvous and docking had been recorded on both of the spacecraft via onboard television and this was transmitted to Soviet receiving centers on the next pass over the Soviet Union. These have not been shown publicly to my knowledge, although the Soviets acknowledged their existence. A few frames have been published in the Soviet press.

Cosmos 186 was recovered on Tuesday, October 31, just one day after the docking event. *Cosmos 188* stayed up until November 2 when it was recovered. The two spacecraft were up for four days and three days, respectively. Again, it appeared that a manned flight was imminent.

Cosmos 212 was launched on Sunday April 14, 1968, at 1000 GMT. The next morning *Cosmos 213* was launched, from Tyuratam also, at 0933 GMT and by 1021 it was docked to *Cosmos 212*. The event was a replay with refinements of the earlier docking in 1967. This one occurred earlier on orbit than had the previous case, for the vehicles were docked by the time they had reached the coast of Chile on South America's western coast. They docked for 3 hours and 50 minutes, separated and continued to perform other experiments and then were both recovered, one day after the other. Each spacecraft stayed up for five days. Hah! now we would get our manned spacecraft.

Summer had come in 1968 and now was on the wane; where on Earth were the expected cosmonauts? The Soviets weren't telling. The obvious thought—it had occurred before—popped into my mind again, that the Soyuz problems with the parachute had given rise to a deep and searching investigation of all of the Soyuz systems in the same manner as the pad fire disaster in the US Apollo program had. I had recognized, time and again, that the popular belief of a low regard for life by the Soviets, so widespread in the US, was not true. For their cosmonauts it was, in fact, quite the opposite. Hence, the long delay in placing another man in orbit.

As a consequence, the radio announcement of the launching of *Cosmos 238* on August 28, 1968, just sounded like one more reconnaissance satellite. All the report contained was that the satellite was in low Earth orbit. That was Wednesday. By the weekend I realized that I was wrong. Some numbers had reached me and they were immediately convincing enough for me to realize that here, still once more, was a disguised Soyuz. To affirm my belief the bird was recovered after only four days in orbit at about 0900 GMT on September 1 in the Karaganda area. Further examination of the orbital data showed that a certain amount of maneuvering had taken place during the orbital sojourn. Practice, I assured myself, makes perfect. I should point out that the maneuvering plus the early recovery just does not fit with their reconnaissance activities.

Some interim excitement was provided when the *Zond 5* spacecraft was launched on September 14, flying on a trajectory that carried it around the Moon and back to a landing in the Indian Ocean on September 22. This was really interesting for much was made of the fact that here was the first water recovery of a Soviet spacecraft. Connection, of course, was immediately made between this flight and a manned, lunar circumferential flight. Great jumpin' balls of fire, had I been on the wrong path after all? Newspaper articles connected the Soyuz command module with the Zond module. Was there any truth in this? Surely there must be some Earth orbit flight of Soyuz before such a lunar attempt. And not just flight but <u>flights</u>! It didn't add up, the *Zond* flight must belong to a separate program.

Even as late as early 1972, there are "stories" that the Zond vehicle consisted of a modified Soyuz command module and a modified service module perhaps, but until the Soviets show the actual spacecraft, the puzzle will remain. Up until the end of 1970 there were three more Zond flights, and nicely keeping step in parallel with them, were the continued stories of manned lunar flight being imminent; so far—December 1975—this has not occurred.

The tales of the big booster, popularly called Webb's Giant during the administration of NASA's James Webb, also pervade the press, magazines, and other writings.

I would like to make it clear that at least for me a big booster, as frequently described, ought to be, and I believe it is, in the works. Why it hasn't flown remains an unanswered question. Even if the explosion-on-the-pad stories are true, more than enough time has passed to recoup those losses and fly the bird. It is completely unlikely that the flaw was so basic as to cancel the project. That sort of decision is utterly out of character for the Soviets; their persistence sets an example for the rest of the world and I do include the United States. I am convinced that the bird is likely to be flown.

I should at this point note an alternate possibility for that booster. The story runs like this. About the time the presumed first flight for the booster was set up (circa 1968) the Shuttle program in the US was getting a good deal of press. A reusable spacecraft is undoubtedly the basis for a continuing space program, manned, or unmanned. The Soviets have a shuttle program. They have always built their spacecraft bigger than those of the US. If the first one or two stages were moderately redesigned—in lieu of the three-stage design—to accommodate the launch of a large recoverable, reusable, winged shuttle, then the launch of such a vehicle would be postponed until the shuttle was ready to fly. The delay, brought about by the introduction of the shuttle, would permit appropriate booster redesign to accommodate the new concept. Both the booster stages and their launch complex would need modification. The big booster described as having twice the thrust of the US Saturn V would then only be used for launching very heavy objects like a space station (or its component sections) or spacecraft to the distant planets, Jupiter, Saturn, etc.

Now came a peculiar event; a Soviet spacecraft was launched at 0900 GMT on Friday, October 25, 1968, but no announcement was made of the name of the spacecraft or of its orbital parameters. However, the next morning the Soviets announced the launching of *Soyuz 3* (at 0834 GMT) and further announced that the day before *Soyuz 2* had been launched—unmanned—and the two spacecraft were in the process of rendezvousing with the manned craft in command of Colonel Georgi T. Beregovoi. Some mild rumors made the press that still a third and manned spacecraft was to be launched. It didn't happen. During rendezvous both spacecraft were in an orbit of 139 by 120 sm, at 51.7°, with an orbital period of 88.6 minutes.

Initially the perigees and period were different but Beregovoi maneuvered his craft on the first orbit to rendezvous within a few feet of *Soyuz 2*. Automatic maneuvering brought the two spacecraft to within 650 feet of each other and then Beregovoi manually operated his ship to complete the rendezvous. I would suppose it reasonable to assume that a docking was intended, however, no discussion of this appeared in the Soviet press so that one can only surmise. As expected, the American press did speculate in this direction but no substantial conclusions were obtained; no revelatory information came to light about any docking attempt—or failure.

Bulletins on Beregovoi were frequent and references to his health were numerous. He was, then, the oldest cosmonaut to be in space and a long-time member of the cosmonaut community. It was also announced on this flight that Soyuz was a multi-compartment spacecraft. Quoting from *Pravda* (27 October 1968) "At the 5th orbit Beregovoi went into a compartment adjacent to the cosmonauts' cabin [i.e., the command module] provided for the performance of scientific experiments." At the end of 16 orbits—one day— Beregovoi remarked, "The flight proceeds normally. No remarks on system operation. Feel good!" No doubt this was a very heartening set of statements to the ground controllers who, obviously, had not forgotten Komarov's problems.

Beregovoi completed a full schedule of experiments and, additionally, put the spacecraft through its paces to firmly establish system reliability and engineering tests that yielded good and desirable results. Confidence was exuded as the mission progressed. A full psychophysiological analysis of Beregovoi and his activities was conducted via spacecraft instrumentation and telemetry. He was observed on television and the TV was broadcast to the public via the Soviet Cosmovision network (spacecraft to ground to public). Seeing as one such picture is worth 100,000 words, it was obvious that Georgi B. was really enjoying himself. Beregovoi was given complete command of his ship and did a bang-up job of running his activities. After separating a great distance from *Soyuz 2* he performed additional rendezvous maneuvers.

Highly placed Academician B. N. Petrov commented in a *Pravda* interview that the prime mission for the flight was "face-to-face" checkout by man without intervening instrumentation for facilities, i.e., with ground control in the loop only in a passive mode. Great emphasis was placed on the adaptability of the highly maneuverable Soyuz to operations with large, permanently manned space stations in Earth orbit.

Soyuz 2 was recovered in a soft landing, after three days of flight, at 0806 GMT in the Karaganda area. This gave the Soviets a preliminary look at the landing procedures for Beregovoi. All was satisfactory.

The newspapers had a field day in two areas. The first encompassed speculation on how long the Beregovoi flight would last—there was really no way of telling. Most commentators and writers looked for two weeks in order to establish a record for spaceflight. The second line of thought, activated at a considerable level—in some publications it was almost a frenzied effort—sought to connect *Soyuz 2* and *3* activities with lunar rendezvous. The flight of *Zond 5* in a circumlunar mission during September 14-22 (1968)—and the later flight of *Zond 6* during November 10-17 (1968)— helped to support these tales of an imminent Soviet manned circumlunar flight. Most writers simply ignored the fact that Soviet philosophy on such matters was predicated by an aura of particular caution. It would take several Soyuz flights in Earth orbit whose length would be equal to or greater than that time required for a circumlunar trip before the Soviets would go circumlunar.

Some of the rumors died in their infancy when Beregovoi landed near Karaganda at a town called Batpak—latitude 50°28' north, longitude

72°41' east—at 0725 GMT on October 30 after the completion of 64 orbits around the Earth. The landing had been planned so well and *Soyuz 3* was controlled so accurately by Beregovoi that the recovery helicopters had landed at the touchdown point minutes before *Soyuz 3* and were there to greet him. His first requests were for an overcoat and hot soup. The temperature was 14 degrees F.

In discussing his flight much information was given. There was not one but two separate liquid propellant rocket engines for the purpose of retro. Each of these could be, and were, used for maneuvering in orbit. The thrust of each engine is 880 lb. The available internal volume is 318 cubic feet (greater than that of the Apollo command module). The orbital compartment has four portholes. One hundred and fifty square feet of solar panels provide 1.3 kilowatts of electrical power. In the retro procedure the selected engine used was ignited and burned for 145 seconds. Following this both the service module and the orbital compartment were separated. The command module then was maneuvered, reentered the atmosphere, and using a lifting reentry method descended through the atmosphere. The "overload" due to acceleration is some 3-4 g's. At 9 kilometers (29,500 feet) a drogue chute opens, followed by the main chute (there is also a backup chute system). When the Soyuz module reaches the ground, at an altitude of three, repeat three, feet, solid rockets ignite reducing the velocity to almost zero and therefore making the landing very soft. A touchdown probe probably serves as the initiating mechanism to ignite the soft-landing rockets.

The Soyuz is a very flexible spacecraft for Earth orbit missions. While on orbit it appears to be capable of supporting three cosmonauts in the command module and three or four more in the orbital compartment, all for a period of five days or so. Two cosmonauts could probably be supported for about a month—this was essentially proven in the flight of *Soyuz 9*. There is access to the spacecraft via a hatch in the orbital compartment for EVA activity.

Later models of Soyuz permit access to other spacecraft without the need for EVA. Television cameras inside and outside the spacecraft are used as desired to observe men and their activities in both areas.

The Soyuz is a large Earth orbiting spacecraft—it weighs about 14,000 lb. It may be scaled up using design modifications for continued future use or, alternately, may be the last nonreusable spacecraft before the multiuse shuttle appears in the Soviet spaceflight arena.

Soviet officialdom was undoubtedly pleased with the results of the Beregovoi mission. He was promoted to Major General and the town of Batpak—his landing site—was renamed Beregovoye in appreciation for his successful flight. Confidence was established in the Soyuz.

With the *Soyuz 2-3* mission having taken place in late 1968, it now appeared that a long stretch of unmanned activities would occur and, perhaps, eight months to a year would pass before another manned flight would take place. This had been the history of Soviet manned spaceflight, and so I turned my attention to other aspects of their program. Besides it was now mid-winter and bitter cold at Tyuratam. To support that last remark the launch of *Venera 5* to Venus on January 5, 1969, was announced as having taken place with the temperature at 30-40 degrees below zero F. Five days later, *Venera 6* was launched under the same temperature conditions. Frigid!

There were some small articles in the press commenting on the presence of the Soviet space-tracking ships *Morzhovets* and *Nevel* in the Gulf of Guinea and the *Cosmonaut Vladimir Komarov* in the northwest Atlantic. Well, I figured that these were connected with the launches of *Venera 5* and *6* and proceeded to set this bit of news back in the files of my memory. I shouldn't have.

On the morning of January 14, I learned that *Soyuz 4* had been launched, again with a single cosmonaut aboard, Vladimir Shatalov.

Launch time was 0739 GMT and the temperatures were still frigid. WOW! I reflected, there's a good bit of launch capability.

Immediately, the rumors flew thick and fast that a second ship would follow with more than one cosmonaut aboard. I needed no convincing this time. I was equally certain that a docking was going to take place. Beregovoi's experiences together with the earlier background provided by *Cosmos 186/188* and *Cosmos 212/213* had been sufficient to permit an important step to ensue; the rendezvous and docking of two manned spacecraft.

Shatalov was initially in an orbit 139.8 by 107.5 sm, at 51°40', with the short period of 88.25 minutes. During his fourth orbit Shatalov changed to 139.4 by 128.2 sm, at the same inclination, and raised the orbital period by one-half minute to 88.75 minutes.

There was no doubt in my mind as to what was about to happen; I bounded out of bed on Wednesday morning (January 15), flipped on the radio, and heard the report that *Soyuz 5* with three cosmonauts had been launched at 0714 GMT. Its orbit, of course, was close to that of *Soyuz 4*. Aboard were Lieutenant Colonel Boris Volynov, Lieutenant Colonel Yevgeny Khrunov, and civilian Engineer Aleksei Yeliseyev. Colonel Shatalov made the observation to ground control that he had visually observed the injection of *Soyuz 5* into orbit.

No docking occurred during the first day but the two spaceships maneuvered extensively, flying in a group flight, practicing rendezvous, testing their radar-finding capabilities, and in general putting both crews and spacecraft through their paces. Numerous experiments and physiological testing of the crews were also carried out during this interval.

Much of the activity aboard both ships was watched via television by the staffs of the coordinating computer center whenever the spacecraft were in range. Some TV transmissions undoubtedly took place between *Soyuz 4* and *5* and the research tracking ships particularly the very well-equipped *Cosmonaut Vladimir Komarov*.

On the fifth orbit of *Soyuz 5*, Commander Volynov maneuvered the spacecraft into a new orbit with apogee at 157 sm and perigee at 131 sm. During orbit 32 for *Soyuz 4*, Shatalov also changed orbit so that his apogee matched that of *Soyuz 5* but perigee was a bit lower at 125 sm. During orbit 33, Shatalov in the active vehicle of the pair—he was flight commander—commenced rendezvous with docking as his objective; the time was 0737 GMT and Shatalov should have been over the narrow waist of South America between Buenos Aires, Argentina, and Santiago, Chile.

At 0820 during their pass over the Soviet Union the two spacecraft, docked; it had taken 43 minutes for the whole procedure. The *Cosmos 212/213* procedures had taken 47 minutes. Communications between the two spacecraft during the final phase of docking was broadcast over Soviet TV and the action published in the American press. Here is an excerpt from the *New York Times* of Friday, January 17 (1969):

> COL. SHATALOV: I am heading straight for the socket.
>
> COL. VOLYNOV: (on *Soyuz 5*) Easy, not so rough!
>
> SHATALOV: It took me quite a while to find you, but now I've got you!
>
> BACKGROUND VOICE ABOARD *SOYUZ 5*: We've been raped, we've been raped!

During the next orbit Khrunov and Yeliseyev having donned their EVA spacesuits proceeded to transfer from the orbital compartment of *Soyuz 5* (which had been depressurized) to that compartment of *Soyuz 4*. They used handrails mounted on both spacecraft and were simultaneously outside their ships for about 1 hour. Naturally, the trip itself did not take that long: the two cosmonauts performed a series of experiments in photography, in mounting and dismounting some equipment, and in some other unspecified tasks.

There was great joy aboard the vehicles, in ground control, and among the Soviet populace especially the Muscovites who are a good bit more sophisticated than some of their Soviet compatriots. The two ships stayed docked for 4 hours and 35 minutes after which they separated at 1255 GMT and continued to conduct a group flight. As usual in a significant advance in space activity congratulations poured in from everywhere. The mission continued to go well.

Toward the end of orbit 48, on January 17, Shatalov deorbited *Soyuz 4* and landed just before 0700 GMT northwest and close to Karaganda; *Soyuz 5* and Volynov remained in orbit. He continued to perform various maneuvers and experiments. The next day near the end of orbit 49 (for *Soyuz 5*) Volynov positioned his spacecraft and fired his retro-rockets. Landing was at 0800 GMT about 125 sm southwest of Kostanay. This city is almost directly north of Tyuratam, the launch base.

After the return of *Soyuz 4* and *5*, all of the cosmonauts were returned to Tyuratam for debriefing and complete medical checkups. The tumult and pleasure of congratulations, greetings by the country's leaders, the cheers of their countrymen and all of the associated honors, pleasures, and banquets awaited the cosmonauts in Moscow. All this took place on Thursday, January 22 (1969) and culminated in a great fireworks display that night together with a 21-gun salute that resounded throughout the city. Moscow had warmly greeted four new "Heroes of the Soviet Union."

If the Soviets had decided to orbit manned spacecraft missions as close as the *Soyuz 2/3* and *Soyuz 4/5* exercises, only a shade more than two months apart, then maybe I could expect still more activity in the calendar year 1969. That would be nice. Midsummer seemed appropriate. But summer brought no manned activity. Another surprise took its place. I was offered a chance to attend the Congress of the International Astronautical Federation to be held in Mar del Plata, Argentina, about 250 sm south of Buenos Aires. This was to be the 20th session of this prestigious organization and I leaped at the opportunity. The conference would be held from October 5 to 10, 1969. It was a propitious choice of month. It was also the first time that this famous international conference was to be held in South America—the Argentina space agency was to be the host.

Since the Soviets were not above taking advantage of such times to pull off a major mission—and rightly so I thought—I was delighted to be in attendance. While I was at it, I decided that Buenos Aires would be a good place for a vacation afterward. It certainly was. The conference stirred a good deal of international interest bringing forth, among others, a large delegation of Russians who came by way of Dakar on the west African coast.

Though the meeting provided a great deal of camaraderie—stirred up still further by the July landing on the Moon of Armstrong and Aldrin—the technical discussions themselves, with few exceptions, were only of minor interest.

I kept getting the impression that members of the Soviet group at the meeting were waiting for something special to occur. This feeling was reinforced by virtue of the fact that a group of cosmonauts that were supposed to be in attendance were not; rumor had it that they were busy at the Tyuratam cosmodrome. The Soviets had on the occasion of previous IAF conferences initiated important spaceflight missions. Sometimes missions were in progress just before the conference started. The last day of the uneventful congress, Friday, October 10, came and went with nary a whisper of any Soviet space activity.

Having come almost 9,000 miles to this conference, I reiterated that a week in Buenos Aires was in order. The seven-hour, 250-mile trip to Buenos Aires via railroad was a trip through miles of lush, flat, emerald-green pastureland covered with endless thousands of Black Angus cattle and interrupted only by picture postcard haciendas. No wonder steak was so cheap here.

The Plaza Hotel in Buenos Aires was a warm, welcome, comfortable sight. Its excellent food and wines and service offer a great contrast to the lesser levels of these items "provided" in New York hotels at greatly exaggerated prices. After luxuriating through breakfast, my wife Elissa and I prepared to do some shopping. This intent was postponed as I picked up an English language morning newspaper and noted headlines blaring the announcement that *Soyuz 6* was in orbit with rookie cosmonauts Georgi Shonin and Valery Kubasov aboard. What's more, the article went on to say unequivocally that additional spaceships were to be launched very soon. The entire article, despite the headline, was rather minimal and I fidgeted the rest of the day trying to get more information. Radio broadcasts were in Spanish or Italian and blabbed endlessly about the local political situations and other trivial—for me—items.

As I walked out of the hotel lobby onto Florida Street, a main shopping area and semi-tourist trap for the unwary purchaser of seemingly pretty goods, who should I meet but one of my Soviet colleagues from the IAF conference? He was warm, effusive, and wore a grin like a Cheshire cat. I congratulated him on the flight of *Soyuz 6;* we exchanged a few words wherein I learned that he was departing for Moscow within a few hours. So, the *Soyuz 6* flight had missed the conference by just one day for here it was October 11, 1969.

Anticipating further action, I swept out of bed on Sunday morning, October 12, and sure enough there was *Soyuz 7* in orbit, with not two but three cosmonauts aboard: Anatoly Filipchenko, Vladislav Volkov, and Viktor Gorbatko. Again, rumors were prevalent that still more launchings were to come. If this were true then the Soviets would be on the verge of establishing still another first; the most manned spaceships in orbit simultaneously together with the most cosmonauts in orbit at the same time. Moreover, stories of multiple dockings and other exotic experiments infiltrated the news thus raising the level of interest and excitement considerably.

Nonetheless, no hard information on details of the mission were available—how I longed for the *New York Times* and *Aviation Week*. The rest of the day passed uneventfully. This really was not a time to be out of the US. On Monday morning we were invited to tea at the home of Professor and Señora Roberto Kugel of the University of La Plata—they had an apartment in Buenos Aires not far from our hotel. Immediately after joining them in their car Roberto told me that one more Soyuz, number 8, was in orbit with two more cosmonauts, Vladimir Shatalov and Aleksei Yeliseyev. That was flight number two for these two cosmonauts. Shatalov was again commander of the mission.

The Soviet Union announced that *Soyuz 7* and *8* had docking apparatus aboard. The implications were obvious but although there was much maneuvering by all three ships and numerous rendezvous exercises no docking ever took place. In view of the large amount of experience represented by the seven cosmonauts in orbit it seemed to me that some sort of equipment failure had occurred to prevent the intended docking. Programmers, control mechanisms, attitude control thrusters, or any one of other electronic or mechanical components could have been the source of the problems. As usual no explanation was ever offered. Attempts to cover the failures by excessive publication of the numerous maneuvering exercises could not hide the fact that all had not gone well.

Aboard *Soyuz 6* the situation was somewhat different for a sophisticated experiment in welding had been carried out inside the orbital compartment—first by depressurizing it and then carrying out a series of welding techniques in the vacuum encompassed in that compartment. The apparatus was called "Vulcan," apt enough, and the experiments were declared quite successful. The experiment was a clue, if somewhat vague, as to future intentions to build large structures in space or to assemble large structures that would be orbited in large sections. Once again, we outsiders would have to wait for future activity.

One thought that did occur to me was that had this whole mission been successful, the Vulcan apparatus might have been used to weld the orbital compartments of *Soyuz 7* and *8* together to form a mini-space station. Since these compartments are discarded at the end of a mission, no dangers would be posed for the occupants of the vehicles involved. Such an experiment, sure as hell, would have been listed as spectacular—much to the delight of the Soviet space planners. But, of course, it did not happen—some other time, perhaps.

The mission set records for the largest number of spacecraft and the greatest number of cosmonauts in orbit simultaneously. Whatever holds had postponed the flights until after the IAF Congress had lessened the general interest, and the orbital problems had made an intended milestone something less than that.

One undeniable item for this mission was the demonstration of the capability to launch three spacecraft, as complex as the Soyuz, in three consecutive days. That was an accomplishment that far exceeds the untrained eye.

The three groups of cosmonauts returned to Earth on October 16, 17, 18, respectively, so that none of the crews spent more than five days in space—the bugaboo of long-term effects of weightlessness obviously was a continuing concern to the Soviet space-medical researchers. As with prior Soyuz flights, these spacecraft used lifting body techniques to fly back through the atmosphere, followed by parachutes, and then touchdown rockets (during the last few feet) to make the landing of the 14,000 lb spacecraft very soft.

Later news releases on the spaceflight troika revealed that a great deal of study of the Earth's surface had been made using a considerable spectrum of photographic equipment in the effort to gather information on the resources and agriculturally bound areas of the Soviet Union. They had tested instrumentation for ascertaining the quality and quantity of the natural resources of the Soviet Union. I expect that studies of crop maturity and health, forest conditions, ice cover, and various aspects of air and water pollution were conducted along with geological investigations. How much, if any, military reconnaissance was conducted from these spacecraft remains a totally unanswered question.

The fanfare following the mission was as expected but didn't have the aura of accomplishment that had accompanied the mission earlier in the year—*Soyuz 4* and *5*. There was an air of "je ne sais quoi" about it all.

I was really bothered by the fact that no Soviet mission had yet flown more than five days—no really long-term capability had been demonstrated. Yet the talk of permanent manned space stations continued to grow and was expanded by and expounded on by Soviet authorities of considerable stature. Surely with all this smoke there must be fire. I now wondered whether I would see launchings in January 1970 as I had in January 1969. The idea didn't seem too unreasonable. If the Soviets were really expanding their mission rates by launching in October 1968 and January 1969 and again in October 1969 then a launch in January 1970 appeared to be in order. It wasn't. Even rumors were at a low level.

The months rolled by and here it was June 1, 1970. Neil Armstrong was in the midst of a visit to the Soviet Union to attend a scientific conference but of course a visit to the Soviet's cosmonaut community was an obvious addition to his agenda. During the course of this visit he addressed the members of this super-exclusive group at a great hall in their community known as Star City, which is on the outskirts of Moscow.

While speaking to the audience of cosmonauts, their wives, and other officials he observed to them that he had placed on the surface of the Moon at Tranquility Base, two medals honoring Yuri Gagarin and Vladimir Komarov. The room grew silent. Armstrong then asked Valentina Gagarina and Valentina Komarova to come to the stage to accept duplicates of these medals honoring their husbands. The only sound in the hall now was the soft weeping of the two women and the tears rolling down the cheeks of Neil Armstrong. The bonds that join these intrepid explorers beyond Earth's atmosphere are known but to few; the rest of us can only stare up at the stars and wonder.

As that day turned to night, Armstrong was invited to the home of Georgi Beregovoi (who had visited the US). While sitting before the proverbial TV set, Georgi turned to Neil and said that here was a flight (unofficially, I suppose) dedicated to him and so saying turned on the set and there was the *Soyuz 9* poised for liftoff. And it did. The first nighttime launching of a Soviet manned flight. Armstrong, of course, was pleased to watch veteran cosmonaut Andrian Nikolayev and civilian rookie Vitaly Sevastyanov liftoff at 1900 GMT. That was midnight at the Tyuratam launch complex.

It took 9 minutes for *Soyuz 9* to reach orbit. This was the first time that an official announcement had been made of the "burn-time" of a Soyuz booster. The time was different from that of the original Vostok booster. It was less.

Nikolayev and Sevastyanov initially went into a low Earth orbit but with, for them, a significant change in perigee. It was 128.6 sm, about 17 miles higher than was usually the case for a piloted ship. Apogee was at 136.7 sm, inclination the usual 51°, and the period was 88.59 minutes.

It was announced that communications would occur on frequencies of 15.008, 18.06, and 20.008 MHz. Moreover, the cosmonauts were able to tune in to frequencies emanating from various countries including the United States *Voice of America* where they heard NASA's congratulations on their flight.

Maneuvers were extensive on the early portion of the flight and perigee was as high as 153.5 sm on the 17th orbit. Additional corrections were made between orbits 49 and 52 and a very finely tuned orbit was obtained during this interval. As a display of their accuracy in being able to locate *Soyuz 9*, ground control released numbers for this orbit interval to three decimal place accuracy most unusual as 162.217 by 150.147 sm, at 51.722°, and having a period of 89.398 minutes. That meant that the location of the spacecraft was known to within 3 feet, a remarkable capability. Expressed another way, the position of the spaceship was known at perigee and apogee to better than four ten-thousandths of 1 percent. This class of measurement capability gives the lie to those hopelessly ignorant humans who are forever making claims about the crudeness of Soviet instrumentation and technological capabilities. Such expertise in one (announced) area certainly implies the same for countless other facets of Soviet industry and science.

The objectives of the flight included much directed toward Earth resources—a continuation of work commenced on the prior three flights—and a wide spectrum of medical studies related to man's ability to withstand the weightless environment.

The official Soviet announcement on the flight stated that it was to be a solitary spacecraft with medical study of man as a prime item.

Nevertheless, the usual rumors, from everywhere—within the Soviet Union, West Germany, Japan, the United States, and all others—playing the guessing game rendered their opinions on the projected length of the flight, its intents, its relation to lunar flight, and so on.

The real and evident clue for a long flight was the raising of the perigee higher than any prior manned flight. This led my compatriots and me to comprise a pool to guess the length of the stay in orbit. Our estimates went from one week all the way to 45 days. As it turned out the first man to join the pool—not me—won by coming within 5 hours of the correct time. At 1335 GMT on June 15 the crew of *Soyuz 9* broke the all-time record aloft set by [Jim] Lovell and [Frank] Borman in *Gemini 7*, several years back [in December 1965]. The *Soyuz 9* crew ignored the event, at least officially. The *New York Times* did not.

The flight was very routine from all the announced aspects of the orbital sojourn. No unusual events took place, no problems, and no Earth-shaking discoveries.

After 424 hours and 59 minutes the *Soyuz 9* crew landed near Karaganda. They had been in orbit for 286 trips around the Earth with a completely successful, if rather routine, program. The flight ended at 1159 GMT on June 19. Aside from readjusting to the gravity field of Mother Earth and some minor biological rhythm adjustments for their bodies, Sevastyanov and the new Major General Nikolayev rejoined their cosmonaut comrades for further studies and new ventures into space. That wasn't to happen for any cosmonaut for almost a year.

The *Soyuz 9* flight brought no early promise of a new more extensive flight. But the length of the flight, almost 18 days, seemed to me to herald the launching of at least a semi-permanent space station in not too many weeks. As it occurred this was not to be the case; a substantial interval passed before the space station was to appear in orbit.

Manned flights were 10 months away—in April 1971. The Soviets managed to keep the space community in a constant state of surprise, nevertheless. For within the next few months the successful flights of *Venera 7*—first spacecraft to actually land on and broadcast from Venus, *Luna 16*—the first unmanned vehicle to scoop up lunar "soil" and return it to the Soviet Union, *Zond 8*—the first spacecraft to circumnavigate the Moon and return to Earth coming in over the North Pole, and *Luna 17*—the carrier vehicle for the most spectacular unmanned exploring vehicle ever on the Moon, *Lunokhod 1*, all took place. It was nice to see the tempo of space exploration kept at a goodly pace. The life blood of science is new knowledge and the Soviets were doing a great deal to keep it flowing in a still new and immeasurably enchanting arena.

So came the morning of April 19, 1971, and from Tyuratam a mighty Zond booster lifted off at 39 minutes after 0100 GMT carrying a new spacecraft that the Soviets called *Salyut 1*. That Monday became really exciting for word had flowed freely, if speculatively, that here was a real dyed-in-the-wool space station; at last! It was placed in a low orbit of about 138 by 124 sm, at 51.6°, and with a period of 88.5 minutes. That was changed soon to a higher orbit. Still no definitive word and still no manned flight.

That situation didn't last long either for at 2354 GMT on Thursday, April 22, off went *Soyuz 10* with cosmonauts Shatalov and Yeliseyev, each for their third flight, and with them rookie Nikolai Rukavishnikov, a civilian engineer. At 0147 GMT on Saturday, April 24, the *Soyuz 10* and the *Salyut* space station docked. The time—after *Soyuz 10* reached orbit and before docking—was spent in orbital maneuvering, inspection of the external areas of the *Salyut*, and to assure the *Soyuz 10* crew that all was working well aboard the *Salyut*. It evidently was…up to the docking exercise.

Though the two spacecraft were docked for 5½ hours the cosmonauts did not get to board the space station. No explanation was offered for this odd turn of events. Some problem obviously did arise—cosmonauts just don't go up to inspect space stations from the outside. Because entry was intended to be an internal transfer—no EVA—I suspect that some failure to pressurize the transfer airlock led to the mission abort. The problem could have been much more serious. A hint that this might have been so is indicated by virtue of the fact that *Soyuz 10* returned to Earth quite soon after undocking.

The total flight time for *Soyuz 10* was only some 47 hours and 46 minutes, less than two days. The landing was safe at 2340 GMT on Saturday, April 24, and, as usual, it occurred near Karaganda. Little discussion of the flight occurred but a hint of things to come became apparent when the orbit of the *Salyut* space station was shifted quite a bit higher; announced on about May 1 as being some 172 by 157 sm. Quite evidently, additional visitors were expected.

The malfunction that prevented the *Soyuz 10* crew from boarding the *Salyut* was due to some problem on the Soyuz spacecraft. It took several weeks to get a good fix on this malfunction. May 1971 went by without any Soyuz flights. The *Salyut* was now in flight for well over several hundred orbits. On a regular basis the Soviet news agency TASS reported on both the orbit parameters and the internal conditions prevailing for the *Salyut* space station. On an as-necessary basis the orbit for *Salyut* was raised to keep it from decaying—that is, to promote stable orbital flight. Internal conditions, as announced, were suitable for supporting human life for an extended period.

The question, of course, was when would this exciting function commence? The month of June rolled around reminding me that a year had passed since the long flight of *Soyuz 9*. When I awoke on Sunday, June 6, the long wait had come to an end; *Soyuz 11* had been orbited at 0455 GMT with Vladislav Volkov (he had been on *Soyuz 7* during October 1969), Lieutenant Colonel Georgi Dobrovolski, and rookie Cosmonaut Viktor Patsayev—a test engineer. Dobrovolski was the flight commander, despite his being a rookie and the overwhelming importance of this flight. Volkov, the youngest of the three, had achieved a great popularity in the eyes of the Soviet public. He was young, 35, handsome, and presented a figure to the Soviet public not unlike that of a popular American folk hero.

Salyut was in about its 775th orbit when *Soyuz 11* was launched. As with the earlier launch of *Soyuz 10* there was some extensive maneuvering and inspection of *Salyut* before the docking procedure was begun on the morning of June 7. During orbit number 789 for *Salyut* (the earlier portion of orbit 19 for *Soyuz 11*) docking occurred at 0745 GMT when both spacecraft were over the Soviet Union. The process was completely successful. Patsayev and Volkov entered *Salyut* first, brought all systems up to working order, reported to ground control on onboard conditions, and only then did Dobrovolski join the other two aboard *Salyut*. Patsayev then returned to *Soyuz 11* and powered it down. Here was a clear indication of preparation for a long flight.

Not only was this to be a long flight but also the business of instituting a many-faceted, intensive, long-awaited experiment program was begun.

Meteorological tests coordinated with the Soviet Meteor weather satellites were used to conduct cloud cover analysis over the Volga valley. Both general cloud cover and cloud structures were investigated directed toward getting a better understanding of weather and learning how—some day—to control that weather. Spectrographic examination of land and ocean from *Salyut* and from airplanes at different altitudes were performed simultaneously in order to correlate the results of these films all taken of the same area.

Various electronic phenomena were studied in order to improve reception of communications between spacecraft and ground control. A number of brand-new devices for studying the reflected light from both planets and stars was put into use. The optical device had the name Orion, either a nickname or an acronym not yet explained to the public. Many atmospheric studies were conducted, other than the cloud cover discussed above, but are not yet fully identified.

Biomedical studies were prolific; ranging from color sensitivity of the eye to investigating atrophy of various muscles of the body in the weightless environment. Despite the fact that one is weightless in orbit, one is not massless, thus contact with the space station walls at any but the gentlest speeds can lead to the same kind of injuries a person might receive on Earth. The problem is certainly more serious in space, what with the paucity of medical aid. As a consequence, the cosmonauts were ordered to wear helmets at all times—at least when moving about—in order to minimize if not obviate head injuries, no matter how slight. No such injuries occurred.

On Thursday, June 24, at 0954 GMT the *Soyuz 11*/*Salyut* crew again set a new record in space travel by exceeding the record set by their compatriots of *Soyuz 9*, Nikolayev and Sevastyanov in June 1970. At this time *Salyut* had completed some 1000 orbits around the Earth while the *Soyuz 11* crew had been up for about 270 orbits. Man was learning to live and work successfully in space. On June 29 at 1828 GMT the *Soyuz 11* separated from the *Salyut* space station. The crew had completed their mission two days earlier and had reentered the *Soyuz* to reactivate the spacecraft in preparation for a return to Earth. Many materials had been loaded aboard *Soyuz 11* that resulted from their voluminous experiments.

Some experiments were left aboard *Salyut*, to be continued, no doubt, by the crew to follow. Final details led to both spacecraft flying together for about three more orbits. At about 2231 GMT, *Soyuz 11* had been placed in its appropriate position for retro firing, which proceeded on schedule. Retro firing takes about 45 minutes. Shortly after firing is complete the service module and the orbital compartment are separated from the command module. The former two modules proceed to reenter uncontrolled and are incinerated in the atmosphere.

Normal procedure for the command module calls for control by the crew so that upon entering the atmosphere a lifting reentry trajectory is programmed to a selected landing site. The spacecraft control is a mixture of automatic programming and manual control by the commander of the craft. The Soyuz is sufficiently sophisticated such that its controls guarantee a very accurate choice of landing site. Moreover, the crew can communicate with ground control during the reentry for all but a few minutes when the hot gas sheath around the spacecraft during the maximum temperature regime blocks out radio signals.

This reentry was different. Almost immediately after separation of the two discardable compartments, long before entering the atmosphere, communications from the crew ceased. All efforts by ground control to raise the voices of the crew were to no avail. Telemetry from the spacecraft continued to be received but no voice communication. It was a frantic and foreboding time for ground control. They had good reason to be worried.

When the orbital compartment was separated from the command module, a pressure equalization valve in the command module failed to open. The cosmonauts were strapped in their seats, all in accord with procedure. The atmosphere in the command module was gone in seconds and the three gallant crewmen died very quickly of air embolism. The nitrogen dissolved in their blood immediately came out of solution creating an ultracritical case of the bends—the "disease" that ocean divers suffer and die from. In addition, all of the atmospheric oxygen that supports life literally blew out of the command module into space.

The actual details on what had occurred was not fully revealed by the Soviet Union until the flight of *Soyuz 13*, over two years following the *Soyuz 11* incident. The failed pressure equalization valve was not supposed to open until well after the spacecraft had entered the Earth's atmosphere but, as noted, it had opened while the *Soyuz 11* was still far in space. In a matter of seconds, the fate of the crew had been sealed, a crew without spacesuits on. There was evidence that the crew had tried to close the valve but the fact that unconsciousness occurs within 10 seconds or so in the event of oxygen depletion and that the crew was strapped down in their seats, foreclosed all efforts to save themselves. Telemetry tapes, read after the accident, showed that the thrust of the escaping air had altered the orientation of the spacecraft and that the guidance system had commanded a readjustment to a correct entry attitude—as it should have.

Soyuz 11 proceeded through the remainder of its reentry; parachutes opened as appropriate, touchdown rockets operated perfectly, and the spacecraft came gently to Earth. The recovery crew opened the *Soyuz* hatch to find the strapped-in crew tranquil in death. It was, and remains, a terrible tragedy. The crew had established a record of 23 days, 18 hours, 383 orbits in space. The cost was a catastrophic one for the Soviets, for the space community, and for all of us seeking new knowledge from areas beyond Earth. The loss of this crew is a never-to-be-forgotten reminder that the exploration of new areas always exacts a toll and the severity of that toll can be great. The *Soyuz 11* crew was buried in the Kremlin wall after cremation and, this time, US astronauts were accepted as official delegates to the funeral services. It was an unhappy day.

Walter Schirra Jr., the former US astronaut, had a column in the *New York Times* of Saturday, July 10, 1971, about the *Soyuz 11* incident.

He expressed the feelings of many of us in relating his regret at the tragedy. A quote from that column is appropriate here:

> We are alike, astronauts and cosmonauts. We are proud of our countries and the work we have done. We are pleased that our two nations are working together in this effort.
>
> Those who have left this planet and returned successfully want to preserve this planet. We know…cabin environment might fail…we work with qualified risks. Man can presume that the environment in *Soyuz 11* failed. THESE THREE MEN WILL NOT BE FORGOTTEN AND MAN WILL CORRECT THE CAUSE OF THAT FAILURE <u>AND WILL CONTINUE HIS QUEST</u>.

The *Salyut* orbit altitude was raised after the return of the *Soyuz 11* crew several times. This led to the belief that, perhaps, the failure could be corrected with relative ease in the defective mechanism. However, as weeks passed and no new Soyuz flight occurred it appeared that the malfunction was very serious, more so than had been initially surmised. Evidently a fix could not be made in a reasonable time for *Salyut* to once again be occupied by a crew. Its supplies were apparently being depleted to the point that support of a new crew did not seem to be either feasible or safe (in the sense of lasting long enough to support a crew for a worthwhile period). In fact, rumor had it that *Soyuz 12* was on the pad when *Soyuz 11* initiated its return. Of course, if that were so, its flight was cancelled.

The *Salyut* was deliberately deorbited by the Soviets on October 11, 1971, over the Pacific Ocean in order that natural decay not occur over land where some of its debris could prove to be a hazard. The appearance of a spacecraft bearing the appellation Salyut was not to occur in Earth orbit for almost 1½ years.

While a design fix might have been a matter of months, the Soviets decided to go over the entire spacecraft looking for any other design discrepancies that might exist. This reexamination extended the interval of no Soyuz flights for over a year and then the next flight was that of an unmanned vehicle testing the redesigned spacecraft.

Cosmos 496 was the label for the spacecraft that was launched on June 26, 1972, just before eight o'clock in the evening—Tyuratam time. The flight lasted six days during which numerous tests were conducted. The returnable portion of the spacecraft was recovered shortly after 1400 GMT on July 2. Rumor has it that a new attempt to launch a Salyut was made in July 1972, but that the attempt failed when one of the booster stages failed to function properly and the spacecraft never reached orbit. Again, firm proof of this event never showed in the open press so one can only speculate based on the thin evidence available. Of course, the intelligence community could enlighten the public but they are not given to talking very much, assuming that they really had something to add in this case. The reluctance of the Soviet Union to offer comment is obvious. Their position just enhances the veil of secrecy that they surround themselves with—if enhance is the correct word.

No further evidence of disguised Soyuz or Salyut spacecraft in Earth orbit can be found and substantiated in the open literature or in other channels up until April 3, 1973, when a spacecraft announced as *Saliut* appeared on the scene. It went up at 0900 GMT and by the next day I had received tracking bulletins indicating that not only the spacecraft but also some 20 odd pieces of "debris" were in orbit with it. Clearly something was amiss for 20 pieces seemed more than normally shed by protective nose cones and the like.

Despite the unidentified objects, the *Saliut* continued in flight and evidently was operational to some degree. On April 8, its orbit was altered by raising the vehicle from a 133 by 161 sm orbit to one of 162 by 184 miles. It had also been raised to an intermediate orbit during April 4-5. The Soviet Union reported daily on the flight, never once intimating that anything was wrong. The spacecraft decayed on May 28 and it was not clear whether or not the decay was natural or had been performed through use of its propulsion system by the Soviet flight controllers.

I concluded that the third stage of the booster had failed to shut down after reaching orbit and had rammed the payload thus damaging it beyond repair insofar as its use by cosmonauts was concerned. I made a number of calculations based upon earlier photos of *Saliut 1* and decided that the objects accompanying *Saliut 2*—especially those that had apogees of up to 400 miles—were gas bottles, either pressurized nitrogen or oxygen, used in the environmental control system and which had broken off from the main vehicle and whose escaping gas had provided enough thrust to place them in the observed orbits. A colleague at the Jet Propulsion Laboratory (in Pasadena, California) made some additional calculations related to the propulsion characteristics of such separated bottles and had arrived at conclusions, which strongly confirmed my own.

Sometime later, a paragraph in the magazine *Aviation Week & Space Technology* suggested that the spacecraft was not a Saliut at all but that the name was used as a disguise to hide the failure of some other new type of spacecraft. Since the name Cosmos seems quite adequate as a cover for new vehicles, one wonders why the Soviet Union would use Saliut, of all names, to disguise any new mission. Such reasoning remains, to say the least, obscure. I, for one, don't believe the *Aviation Week* tale.

Before *Saliut 2* said its goodbyes to Earth orbit, *Cosmos 557* was launched on May 11 into a 135 by 165 mile orbit at an inclination of 51.6°; nominally the angle used for manned or man-related flights. The guessing games commenced again; it was a new Saliut (hardly likely), it was another unmanned Soyuz, it was a lunar landing module on a practice mission.[74] It is quite reasonable to suppose that this was a man-related mission since the usual Soviet tracking ships were on station at sea in conjunction with the flight—or were they still watching *Saliut 2*? Observers found that *Cosmos 557* did not perform any maneuvers in orbit. That was rather unlike man-related spacecraft. Again it appeared that all had not gone as it should. The final word was that the vehicle had decayed on May 22 and descended into the ocean at about 0300 GMT at 41° south, 122° east some hundreds of miles south of Esperance, Australia.

If *Cosmos 557* were indeed another Saliut, it would imply that the fix for *Saliut 2* was a simple one and that, moreover, the Soviets had these expensive and complex space stations on some sort of "mass" production line. How else could another one be launched (with adjustments to account for the failure of *Saliut 2*) as quickly as had occurred? Both such implications do not appear correct.

First, the Soviets have not been known to respond that soon to any kind of failure since the very earliest days of their space program when we were all very naive about the real sophistication required for success. Second, to assume that the Saliut was in serial production seems to be stretching matters involving multi-hundred million dollar (ruble) spacecraft. While it is possible that *Cosmos 557* was a boiler-plate test of a man-related bird, one plausible explanation, the specific intent remains a mystery that may be revealed at some future time...should the Soviets so choose.

One might speculate—again—endlessly as to why a large and powerful nation like the Soviet Union is unable to publicly admit such mishaps and take them in stride but unearthing the Soviet psyche

---

74 In his book *Carrying the Fire: An Astronaut's Journey* (Farrar, Straus and Giroux, 1974), Mike Collins tells of a conversation with Cosmonaut Pavel Belyayev (at Le Bourget in 1967) in which Belyayev states he expects to fly around the Moon before long. Hence, a lunar module test is not out of the question in the trouble-plagued Soviet lunar (manned) program.

means digging back to czarist characteristics of secrecy and their effects on the nature of present Soviet society and politics. I leave that to other writers for it is a subject covering a vast expanse in itself.

While the jigsaw puzzle of *Saliut 2* and *Cosmos 557* lay scattered about the landscape unsolved, untidy, and unsettling yet another man-related spacecraft went into Earth orbit on June 15, 1973, quite early in the morning—at 0558 GMT. It was recovered just 32 orbits later at 0553 GMT on June 17. The Soviet Union seemed to have further questions on the redesigned Soyuz—or so it appeared. While so short a flight offered the distinct possibility that a new generation of spacecraft was being tested, speculation focused on a Soyuz. If a new bird was in the offing, as of March 1974 it had not made another appearance. So, *Cosmos 573* was accepted as a Soyuz and it fitted the concept of endless and supercautious testing of any and every new change in a manned spacecraft. Obviously, better safe than sorry.

At long last on September 27, 1973, at 1218 GMT, *Soyuz 12* with Cosmonauts Oleg Makarov and Lieutenant Colonel Vasily Lazarev was launched from Tyuratam. For the first time the Soviet Union actually announced beforehand the intended length of the flight... two days. Again, a test flight to check out the spacecraft, but with confidence now high enough to place men aboard. Good! There were only two men aboard because of the bulky spacesuits that they were wearing as extra protection against any kind of depressurization event. Two men so dressed precluded the presence of a third man by virtue of the lack of room in the cabin. *Soyuz 12* entered a 121 by 155 mile orbit and soon transferred to a 202 by 214 mile orbit where the spacecraft was put through its paces; where some astronomical and other research was carried out. This was a test of compatibility between man and a "new" vehicle, no more, no less.

Promptly, as scheduled, Lazarev and Makarov returned to Earth at 1134 GMT on September 29 after a flight of 47 hours and 16 minutes. Comments in the world press were minimal, but the TASS communiques indicated satisfaction with the flight. Nowhere did any discrepancies appear, so that the test was assuredly a success. The puzzling question was, what would happen next? A Saliut flight, with *Soyuz 13* and *14* docking with consecutive crews...or what? *Soyuz 13* made its appearance on Tuesday, December 18, 1973, at 1154 GMT. Its two-man crew was Major Pyotr Klimuk and civilian Engineer Valentin Lebedev.

No announcement of the flight duration was forthcoming this time although various maneuvers were conducted and announced rather routinely. This was the first time in the history of manned spaceflight that crews of the Soviet Union and the United States were in orbit simultaneously. *Skylab* with its third crew was in orbit. This situation remained so for the next eight days. The two-man crew conducted Earth resources investigations and a number of astronomy experiments. With a satisfactory checkout of the spacecraft and crew having been completed, the return to Earth was made at 0850 GMT on December 26. The recovery area being in the vicinity of Karaganda, a steel and coal mining center some 550 miles northeast of Tyuratam. It should be noted that the Karaganda area has been the standard recovery zone for Soviet cosmonauts ever since the flight of *Vostok 3* in August 1962.

While the Soviet Union launched 86 spacecraft into or beyond Earth orbit in 1973, their manned spaceflight program had barely begun to show signs of revival and had shown some great difficulty insofar as the space station aspects of it were concerned. It remained to be seen if the year 1974 would show new strengths and sufficient success to bring their long-range goal of a large, permanent, manned facility in Earth orbit closer to reality.

The year 1974 opened with the, by now, standard slower rate of satellite launchings for the earlier part of the year; nothing eye-catching happened until mid-February when four Mars spacecraft commenced arrival at Mars.[75] By mid-March all had arrived and all had essentially failed in their missions. Three missed the planet. The one successful ejection of an intended surface lander from the mother-bus broadcast only during descent into the Martian atmosphere, but never relayed information from the surface. So, $1.2 billion in hardware yielded almost no results.

At the end of March, the Soviet Union successfully placed *Cosmos 637* in synchronous orbit. By virtue of its identification—a test bird rather than an operational one. On May 29, *Luna 22*, a lunar orbiter, was launched and shortly thereafter, on June 3, reached lunar orbit and proceeded to follow up the program commenced by *Luna 19*.

Finally, on Monday, June 24, at 2237 GMT, the event I was waiting for occurred; *Saliut 3* was launched into a low Earth orbit and quickly raised to a higher, more stable orbit. No doubt, to await a crew via the next Soyuz spacecraft. After a nine-day checkout of the *Saliut 3* space station, up came *Soyuz 14* with cosmonauts Colonel Pavel Popovich and Colonel Engineer Yuri Artyukhin aboard. Popovich's presence certainly was a surprise for he had not flown since *Vostok 4* way back in August 1962, 12 years earlier.

There was no unusual activity during the flight; most experiments were in the medical sphere. However, there was some indication that *Saliut 3* carried very large optics, which thereby had a military implication. If true, *Saliut 3* could easily be a Soviet version of the scrapped US MOL program (Manned Orbiting Laboratory, a euphemism for a manned reconnaissance spacecraft).

No untoward incidents were reported and the *Soyuz 14* crew landed near Karaganda after a 16-day flight. They came down shortly after

---

75   They had been launched from Tyuratam during July and August 1973.

1200 GMT on July 19. Since the time zone of Karaganda is some 5 hours later than GMT, the landing was made, locally, in the late afternoon.

It seemed to me that another crew should replace that of *Soyuz 14* aboard the *Saliut* space station within two or three weeks. That estimate proved to be a bit optimistic but on August 26 at 1958 GMT, *Soyuz 15* was launched from Tyuratam with two rookie cosmonauts aboard; Lieutenant Colonel Gennady Sarafanov and Colonel Lev Demin. The latter was the oldest crew member to fly a first flight, being a grandfather at age 48. It seemed, however slowly, that the population of space crewmembers was no longer concentrated in the young.

The liftoff time raised many a quizzical eyebrow, including my own, for it was in the early morning hours at Tyuratam in quite thorough darkness. It seemed reasonable to assume that the length of the flight would be long; adjusted so that the return to Earth would occur during daylight hours. I recalled that the flight of *Soyuz 9* also was initiated with a nighttime launch. So, there was a precedent.

I expected another uneventful flight. It was not to be. The Soviet manned spaceflight program still does not take anywhere near full advantage of man's intrinsic capabilities. This spacecraft was automated to the extent of having the rendezvous and docking performed without crew participation. Not only without crew participation but also, from all the evidence, with no plan for manual completion of the task if the automated functions did not come off as planned. They did not go as planned.

The crew had to make an emergency landing two days later at 2010 GMT, southwest of Tselinograd, which is northwest of and close to Karaganda. At 2010 GMT means a landing at just after 0100 local time, in the wee hours of the morning, in thorough darkness—just like the launching. Present-day spacecraft do not have landing lights and landing under those conditions leaves something to be desired. While the return turned out to be perfectly safe, the crew was awarded not only the usual Hero of the Soviet Union but also the Order of Lenin medal as well. Since the Lenin award is only given out with great care, I surmise that there is a goodly portion of that flight experience yet to be told.

Sufficient concern was raised about the flight such that in the US, a number of senators questioned the readiness of the Soyuz spacecraft for the joint Soviet-American flight scheduled for July 1975.

General Vladimir Shatalov, a veteran of three Soyuz flights (*Soyuz 4, 8,* and *10*) and now the Chief of Cosmonaut Training attempted to alleviate concern in the course of several statements to the press made in Moscow, at the cosmonaut training center (Zvezdny Gorodok—Star City) and at the NASA Johnson Space Center near Houston, Texas.

To date all of his remarks appear to be in vain. Shatalov has implied a failure in the spacecraft without specifically saying it in those words. He tried to convince the public that the night landing was intended and that the actions of the automated spacecraft were almost as prescribed beforehand. That...is too difficult to swallow whole. It clearly implies that the Soviet Union has produced so many Soyuz spacecraft that they can easily afford to expend them in two-day flights, that it does not matter if a crew fails to dock with, board, and use the expensive Saliut in orbit, and that the crews of such automated vehicles are at the mercy of their spacecraft. Bunk!

Should he really be serious about such a philosophy then he needs some thorough reeducation. There is no machine anywhere that can reprogram, judge, reason, reject, perceive, and select, like a man can. I've said it before and repeat again, no such machine will be built on Earth for one hell of a long time especially keeping in mind the weight, volume, and versatility ramifications involved; even for relatively simple operations. I certainly would very much like to see great success in their manned space station program. It will not come off unless man can participate in many, many, more functions than permitted under the philosophy that rules the present program.

It appeared that the failure of *Soyuz 15* to dock with *Salyut 31* would once again lead to a long interval commensurate with a searching analysis of the *Soyuz 15* problem before any *Soyuz 16* was launched. While four months did pass before the event, I had actually supposed it would be much longer in keeping with the ultracautious progress of the Soviet space program.

Soyuz 16 was launched into orbit on December 2, 1974, at 0939 GMT. On December 3, the Soviet Union announced that this was a full-dress rehearsal for the upcoming Apollo-Soyuz flight scheduled for July 1975. It was evident that the Salyut station would not play any significant part in this flight. Cosmonauts Anatoly Filipchenko, who had flown in *Soyuz 7*, and Nikolai Rukavishnikov, of *Soyuz 10*, were scheduled for a short time on orbit. *Soyuz 16* stayed in a relatively low 140 by 110 sm orbit, at an inclination of 51.8°, slightly different from the 51.6° of *Salyut 3* and *Soyuz 15*.

Various exercises were conducted to simulate the actions that were to occur in the Apollo-Soyuz flight. A number of *Soyuz 16* systems had been revised to suit the needs of the future flight. In addition, Earth surface photography and other observations were made and a series of experiments were performed, again, as practice for the July 1975 international effort. The cabin atmosphere was selected as 10 psi; the same as that selected for the Apollo-Soyuz event. The entire tracking network of the Soviet Union was intently engaged in the flight and the NASA network of the US was also given information for tracking purposes to lend support for the scheduled joint enterprise. The Soviet Union also used the flight to aid in Earth resources studies; a further indication of their previously stated intent to make substantial use of man in that area.

The joint flight was to use an entirely new docking mechanism where for the first time an androgynous system would be used so that any pair of spacecraft so constructed could dock. These docking processes were simulated using test hardware. All the tests were eminently successful. The crew circularized the orbit to the one to be used in the joint flight but, before descent, they switched to an elliptical one more suitable for their reentry procedures.

A collection of biological specimens was carried aboard; among them were fast-growing plants and ring fungi to permit observation of growth phenomena in zero g. Fish eggs, which developed to young on orbit, permitted observation of their behavior (and then back on the ground their physical development) resulting from the zero-gravity environment.

Having successfully completed the hardware studies related to the future flight, the Earth resources photography and associated experiments for near-term usage and the biological experiments whose results would shed light on growth and other life processes in zero gravity, the *Soyuz 16* returned to Earth, touching down 185 sm north of the city of Dzhezkazgan at 0804 GMT on December 8. The flight had lasted 142 hours 25 minutes, encompassing 96 orbits of the Earth. It was interesting to note that, for the first time since 1969, three flights had taken place during 1974; a certain indication of improved reliability and confidence in the activities of the Soviet Union's manned space program.

As 1974 drew to a close, I mused on the fact that new manned spaceflights might be as far away as the spring of 1975—for winters in the Soviet Union are fierce with temperatures frequently well below zero for long periods and often accompanied by large amounts of wind-driven snow—blizzards. While Tyuratam is in the southern

Soviet Union, its latitude is still farther north than that of almost all of the United States. Although *Soyuz 4* and *5* were launched during subzero conditions in January 1969, that event seemed to be an exception and I did not expect a recurrence.

Those surmises were blown away by the winter winds when *Salyut 4* was launched at 0414 GMT on December 26, 1974. A careful checkout of systems would take two weeks; hence occupancy should then occur in mid-January. It was to be sooner. From an initial orbit at 51.6° and 168 by 136 sm, the Salyut was soon raised to 221 by 213 sm—still at 51.6° but with a period of 91.3 minutes. That orbit implied a long operational life for it was higher than either *Salyut 1* or *3* had been. Conditions aboard *Salyut 4* were reported to be normal in bulletins issued over the following two weeks.

The first launching of 1975 occurred at 2143 GMT on January 10. *Soyuz 17* was placed into an initial orbit of 155 by 116 sm, followed quickly by a change to 220 by 182 sm. In the early morning of January 12, a further orbit correction was made and Cosmonauts Alexsei Gubarev and Georgy Grechko docked with the space station, *Salyut 4*. A long series of complex medical/biological experiments were soon initiated, all aimed at deciphering man's many reactions to the weightless environment. Heart operation, venal blood circulation, blood vessel tone, blood content of the brain and its changes early in the flight, and negative lower body pressure response constituted the first experiments during the flight. This was a positive indication of the detailed intent of the medical studies to be conducted over the length of the mission. While predicting the length of orbital flights in the manned area is hazardous at best, this one gave the appearances of a long stay. I guesstimated a month or more.

Because the *Salyut 4 / Soyuz 17* station was in an orbit promising a long life there was much speculation that several crews would occupy the station through the year thus leading to a situation where it might be occupied during the Apollo-Soyuz mission. And, indeed, that is exactly what happened.

The two 43-year-old cosmonauts, who were at the time space rookies, televised pictures to Earth regularly—all of which, incidentally, had great clarity and a further improvement over earlier video from space. Voice communications was conducted on a frequency of 121.75 MHz and telemetry via a frequency of 922.75 MHz. Astronomical tasks using both optical and X-ray telescopes comprised only a portion of the scientific array of work in their schedule.

News of the experiments being conducted aboard *Salyut 4* appeared regularly in *Pravda* with frequently voluminous details as though prepared for publication in a technical journal. Moreover, regular reports were forthcoming on the cosmonauts' state of health. Finally, on February 9 at 1103 GMT, *Soyuz 17* came back to Earth 70 miles northeast of Tselinograd. The cosmonauts landed in 35 knot winds under conditions of poor visibility. They had actually separated from *Salyut 4* about 5 hours earlier. Before they left it, the station was made orderly and systems placed in quiescent conditions to await the arrival of a new crew, an event certain to occur.

At long last the Soviet Union had surpassed its own flight record of 23 days set by the *Soyuz 11* crew. The 30-day flight was two days longer than the American *Skylab 1* flight. Although to capture the world record for spaceflight, both the *Skylab 2* flight of 59 days and the *Skylab 3* flight of 84 days would need to be surpassed. The gap was still very substantial; the real question was, how long would the gap last? Only the passage of time would tell.

It was evident that a very, careful, incisive examination of both the *Soyuz 17* cosmonauts and the results of their numerous experiments were to be conducted before a new flight would ensue. The likelihood of a *Soyuz 18* following the latest flight, as quickly as *Soyuz 17* followed *Soyuz 16*—only 33 days—did not seem to be in the realm of the expected.

Meanwhile, the orbit of *Salyut 4* was renewed several times to overcome the effects of drag and these events were announced.

Nevertheless, less than two months after the return of *Soyuz 17*, on April 5, a Saturday, Vasily Lazarev and Oleg Makarov, the cosmonauts of *Soyuz 12*, were launched from Tyuratam aboard what would have been designated *Soyuz 18*. For the first time in their manned space program, a Soviet booster failed. The mishap was caused by a malfunction of the latch mechanism connecting the sustainer stage and the third stage. Half of the set of latches did not disconnect and thus dragged along the expended sustainer. Third-stage ignition actually took place and lasted 4 seconds. When the stage separation failed to occur properly, the vehicle deviated radically from its prescribed trajectory. That generated a signal for the upper stage to shut down and for the manned command module to be separated from the rest of the spacecraft and booster through operation of the escape system.

The escape system worked very well and the cosmonauts landed down range about 1,000 miles from Tyuratam southwest of the city of Gorno-Altaysk in western Siberia. During the course of the escape operation, the cosmonauts expressed their concern, in really strong terms, about the possibility of landing in China; an event they desired to avoid at all costs. Communications were not as good as they could have been and the rescue team did not reach the cosmonauts for almost a day in the mountainous region where they landed. Through whatever communications they did have, Makarov and Lazarev, frustrated as they were, raised unbridled hell with ground control. One readily sympathizes with them in so uncomfortable a situation as well as with ground control who were totally concerned with completing an effective, safe rescue. Suffice to say, all was successful in the rescue operations.

One hundred and four days after Grechko and Gubarev left *Salyut 4*, Pyotr Klimuk and Vitaly Sevastyanov, both veterans of earlier flights, lifted off at 1458 GMT on May 24 and went into a low Earth orbit almost identical to the initial path for *Soyuz 17*. The new spacecraft was called *Soyuz 18* (the April 5 flight was left unnumbered and simply called an anomaly) and was raised to match the orbit of *Salyut 4*. Rendezvous and docking were followed in short order by the boarding onto the station. It was now possible that the challenge to the spaceflight record would be made but, alternately, a more cautious approach might prevail that called for doubling the previous flight period as a next goal. The Soviet intent remained to be seen.

The *New York Times* of May 26, 1975, in an article discussing early phases of the *Soyuz 18* flight carried remarks attributed to "western sources" stating that this flight was quite unexpected since it had been presumed *Salyut 4* would be abandoned, like its predecessor stations, after the *Soyuz 17* flight. It is worth noting that an observation on the higher orbit for *Salyut 4* over those of its antecedents, which clearly implied a longer lifetime, was absent from the remarks. What was observed was the fact that the Apollo-Soyuz flight was only seven weeks away and the likelihood of three manned spacecraft in orbit simultaneously was all but assured.

Early experiments covered planetary and stellar astronomy, examination of specific radiation from the Sun, and a group of medically oriented studies. Preceding all experiment work, however, was checkout and verification of major subsystems aboard *Salyut*: life support, power supply, and heat regulation. Camera film supplies were replenished and various biological specimens were placed in their proper modules. Evidently, new specimens had been brought up to the station aboard *Soyuz 18*. An interesting experiment in laser tracking *Salyut* from the ground was also performed—the station had a corner reflector array attached to its surface, which reflected laser beam signals sent from Earth stations.

A controlled garden called "Oasis" was aboard and initially onions and peas were grown and studied. A considerable segment of time was allotted to Earth surface photography; it had been announced that millions of square kilometers of the Soviet Union were to be photographed in connection with Earth resources studies, "for national economic needs." Experiments in the realm of physics were as sophisticated as, for example, determining the isotopic and chemical composition of galactic and solar cosmic rays and radiation characteristics of X-ray sources in the constellations Scorpio, Virgo, and Cygnus.

Continual bulletins on the physical state of the cosmonauts assured the public that they were in good health. Special experiments had been devised to stress the muscle structures of the body to counteract the absence of gravity. Connected to a spectrum of medical instrumentation each cosmonaut, in turn, rode the bicycle ergometer for cardiovascular testing. A perusal of the experiments and studies conducted reveals that a wide array of activity was included covering the immediate health of the crew, a specific series of medical and biological studies, a number of fundamental investigations in the arena of astrophysics, and, quite important, detailed engineering studies on specific operation of the station itself and all its subsystems including those which permitted autonomous operation in the unmanned mode. Clearly Cosmonauts Klimuk and Sevastyanov did not have much free time.

Also, aboard *Salyut* was an automatic computer-controlled examination/readout device for determining all orbit parameters for the Salyut-Soyuz spacecraft. The device was termed the Delta navigation system. On July 15, the two *Soyuz 18* cosmonauts found that they had orbital companions in *Soyuz 19*, launched at 1220 GMT as the first half of the Apollo-Soyuz international mission. Alexei Leonov and Valery Kubasov were to join the crew of an Apollo spacecraft in an orbit about 80 miles lower than that of the *Salyut*. The Apollo crew of Thomas Stafford, Vance Brand, and Donald [Deke]

Slayton left the Kennedy Space Center at 1950 GMT on their way to the historic rendezvous and docking with *Soyuz 19*.

Now there were four spacecraft in orbit; three Soviet and one from the United States, and four cosmonauts and three astronauts were aboard those four craft. The distinctly different orbits for the two groups of spacecraft assured that there would be no problems in traffic control. Since all the Soviet craft used the same frequencies, various exchanges of greetings and mutual good wishes took place. There was at least one lengthy conversation among the cosmonauts aboard *Salyut* and *Soyuz 19* before the Apollo-Soyuz rendezvous and docking (rendock) took place. The substance of a part of that was that all the world was watching hence great care should be exercised so that all went well. It was and it did!

The month of July was largely over, *Soyuz 19* had returned to Earth and the crew of *Soyuz 18* now having completed the multitudinous investigations prepared themselves and both spacecraft for the trip back to the Earth's surface. The *Soyuz 18* retro-engine was fired in a brief test and its other systems were checked out. Once again *Salyut 4* was set up for autonomous operation. At 1056 GMT on July 26, *Soyuz 18* undocked from *Salyut 4*. Some 3 hours later along their orbital path the retro-engine was ignited for a full burn. At 1418 GMT the command module made a soft landing 35 miles northeast of the town of Arkalyk in Kazakhstan. The flight had covered 63 days, more than twice the length of the *Soyuz 17* orbital sojourn. A long successful flight covering 1511 hours and 994 orbits of the Earth was now part of history.

A brief review of the flight appeared in *Pravda* the day after the landing and included promises of "more complex space investigation problems" and "the creation of permanently functioning long range orbiting stations in space."

The *Salyut 4* continued to operate in its unmanned mode with a number of systems returning engineering information as well as other data long past the end of 1975. This continued even after the *Salyut 5* station was orbited and occupied in mid-1976. *Soyuz 19* had originally climbed into its orbit where a number of checkout and housekeeping tasks were completed; thereafter the crew got a good night's sleep. On July 16, during their 17th orbit, they circularized it to 140 by 138 sm to await the mission rendezvous. Since the Apollo had been designed for lunar flights, it had a much larger propellant capacity than Soyuz, so it had been chosen as the active vehicle for the rendock exercise.

During their 20th orbit, Leonov and Kubasov sent back the first video of the flight. At liftoff it was discovered that their video camera was not working. In the interim the cosmonauts had, with the aid of ground control, made repairs. On July 17 at 1612 GMT Apollo and *Soyuz 19* docked using the US-built docking module and the androgynous docking hardware built by both countries for their respective spacecraft. After the successful docking—as stated in *Pravda* on July 18—"In conformity with the agreements between the Soviets and Americans the cosmonauts will exchange flags, texts of the agreement between the USSR and the USA on cooperation in the research and use of outer space for peaceful purposes, memorial medals and plaques as well as seeds of trees which will be planted in the USSR and the USA. The crews will witness the execution of the docking of the *Soyuz 19* and *Apollo* on the docked crafts."

After mutual congratulations, posing for photographs in mixed crews, exchange of crews from one spacecraft to the other, messages from Chairman [Leonid] Brezhnev and President [Gerald] Ford, an exchange of national flags, signatures of documents proclaiming the first international crew (of Earth), passage of a United Nations (UN) flag from the cosmonauts to the astronauts (for later presentation to the UN), and an exchange of instruments for experiments, the crews settled down to conduct a number of biological and materials/mechanics experiments.

Of course, the crews held a joint press conference commencing at 1730 GMT when the ships were over the Poland/Soviet border. The conference lasted for 1½ hours. During the docked portion of the flight the internal pressure of both craft was maintained at 10 psi; that meant raising it 5 psi from the standard atmosphere aboard *Apollo* and lowering it 5 psi from the atmosphere usually maintained on *Soyuz* spacecraft.

At 1202 GMT on July 19, the two craft undocked and further experiments were conducted using the *Apollo* to eclipse the Sun and therein act as a coronagraph for photography conducted from *Soyuz 19*. A second docking occurred at 1240 GMT, with *Soyuz 19* as the active craft, and at 1326 the final undocking occurred and two spacecraft separated. The Soyuz crew then completed scheduled Earth resources photography, carried out some additional experiments, and made early preparations for reentry to occur on July 21. *Apollo* was to stay aloft for nine days and had scheduled its own protocol for further investigations in orbit.

Cosmonauts Leonov and Kubasov landed at 1051 GMT on July 21, after 142 hours 31 minutes of flight, near the town of Arkalyk where their *Soyuz 18* compatriots were to land five days later. The *Apollo* crew performed their usual mid-ocean landing safely on July 24 at 2118 GMT.

The pace of Soviet manned spaceflight activity during 1975 had led to a substantial addition of man-hours aloft. All the evidence pointed toward a continuation of such activity into the year 1976; perhaps, one might even anticipate a further increase in Soviet manned spaceflight duration.

Table VIII: Launches of the Spacecraft Soyuz

Soyuz No.	Crew	Launch Date Day GMT	Ha / Hp	Inc.	Period	Recovery Date	Days Aloft
1	Komarov	4/23/1967 Sun 0035	139.2/124.9	51 40'	88.6	4/24/1967	1
2	unmanned	10/25/1968 Fri 0859	139.2/115	51.7	88.5	10/28/1968	3
3	Beregovoi	10/26/1968 Sat 0834	139.8/127.4	51.7	88.6	10/30/1968	4
4	Shatalov	1/14/1969 Tue 0739	139.8/107.5	51 40'	88.25	1/17/1969	3
5	Khrunov Volynov Yeliseyev	1/15/1969 Wed 0714	142.9/124.3	51 40'	88.7	1/18/1969	3
6	Kubasov Shonin	10/11/1969 Sat 1108	138.6/115.6	51.7	88.36	10/16/1969	5
7	Filipchenko Gorbatko Volkov	10/12/1969 Sun 1044	140.4/128.6	51.7	88.6	10/17/1969	5
8	Shatalov Yeliseyev	10/13/1969 Mon 1029	138.6/127.4	51.7	88.6	10/18/1969	5
9	Nikolayev Sevastyanov	6/1/1970 Mon 1900	136.7/128.6	51.7	88.59	6/19/1970	18
10	Shatalov Yeliseyev Rukavishnikov	4/22/1971 Thu 2354	154/129.8	51.6	89	4/21/1971	2
11	Dobrovolski Volkov Patsayev	6/6/1971 Sun 0455	135/115	51.6	88.42	6/29/1971	24

Chapter 8: Soyuz

Soyuz No.	Crew	Launch Date Day GMT	Ha / Hp	Inc.	Period	Recovery Date	Days Aloft
12	Lazarev Makarov	9/27/1973 Thu 1218	154.7/120.5	51.6	88.6	9/29/1973	2
13	Klimuk Lebedev	12/18/1973 Tue 1154	169/139.8	51.6	89.22	12/26/1973	8
14	Popovich Artyukhin	7/3/1974 Wed 1851	172/158.5	51.6	89.7	7/19/1974	16
15	Sarafanov Demin	8/26/1974 Mon 1957	145.8/106.4	51.57	88.49	8/28/1974	2
16	Filipchenko Rukavishnikov	12/2/1974 Mon 0939	138.6/10	51.8	88.4	12/8/1974	6
17	Gubarev Grechko	1/10/1975 Fri 2143	155/116	51.6	88.4	2/9/1975	30
18	Klimuk Sevastyanov	5/24/1975 Sat 1458	153.5/119.9	51.6	88.6	7/26/1975	63
19	Leonov Kubasov	7/15/1975 Tue 1220	149/115	51.76	88.68	7/21/1975	6
20	unmanned	11/17/1975 Mon 1436	163.7/124	51.6	88.8	2/16/1976	91
21	Volynov Zholobov	7/6/1976 Tue 1209	170.2/152.7	51.6	89.65	8/24/1976	50
22	Bykovsky Aksenov	9/15/1976 Wed 0948	183.6/114.3	64.75	89.3	9/23/1976	8
23	Rozhdestvensky Zudov	10/14/1976 Thu 1740	170.9/151	51.6	89.5	10/16/1976	2

Apogee, perigee in sm, inclination in degrees and minutes, period in minutes. Days may not be whole days. All orbits are the earliest to be determined from records and raw data; in some cases—*Soyuz 14*—there may have been a lower perigee.

Table IX: For 1967-1975, Flight (Spacecraft) Hours and Man-Hours for the Flights of the Soyuz Spacecraft

Flight Date	Soyuz No.	No. Orbits	Total Orbits to Date	Flight Hours	Flight Hours to Date	No. in Crew	Man-Hours	Man-Hours Total to Date
4/23/1967	1	18	18	26:42	26:42	1	26.7	26.7
10/26/1968	3	64	82	94:51	121:33	1	94.85	121.33
1/14/1969	4	48	130	71:14	192:47	1	51.85	173.4
1/15/1969	5	49	179	72:46	265:33	3	58.15	231.55
						3	84.8	316.35
						1	44.5	360.85
10/11/1969	6	80	259	118:44	384:17	2	237.46	598.31
10/12/1969	7	80	339	118:42	502:59	3	356.1	954.41
10/13/1969	8	80	419	118:41	621:40	2	237.36	1191.77
6/1/1970	9	285	704	424:59	1046:39	2	851.97	2043.74
4/22/1971	10	32	736	47:46	1094:25	3	143.3	2187.04
6/6/1971	11	382	1118	570:01	1664:26	3	1710.05	3897.09
9/27/1973	12	31	1149	47:16	1711:42	2	94.52	3991.61
12/18/1973	13	127	1276	188:56	1900:38	2	377.86	4369.47
7/3/1974	14	251	1527	377:30	2278:08	2	755	5124.47
8/26/1974	15	32	1559	48:13	2326:21	2	96.13	5220.6
12/2/1974	16	96	1655	142:25	2468:46	2	284.83	5505.43
1/10/1975	17	465	2120	709:20	3178:06	2	1418.67	6924.1
5/24/1975	18	994	3114	1511:20	4689:26	2	3022.67	9946.77
7/15/1975	19	96	3210	142:31	4832:57	2	285.07	10,231.84

Flight hours and total in hours and minutes; other times in hours.

Table X: Tests of Soyuz Spacecraft under the Cosmos Appellation

Cosmos No.	Launch Date/Day/ GMT	Ha	Hp	Inc.	Period	Recovery Date	No. Orbits	Days Aloft
133	11-28-66 Mon 1100	144	112.5	51.9	88.4	11/30/1966	32	2
140	2-7-67 Tue 0320	149.8	105.6	51.7	88.48	2/9/1967	32	2
186	10-27-67 Fri 0930	146	130	51.7	88.7	10/31/1967	64	4
188	10-30-67 Mon 0810	171.5	124.3	51.7	88.97	11/2/1967	49	3
212	4-14-68 Sun 0959	148.5	130.5	51.7	88.75	4/19/1968	80	5
213	4-15-68 Mon 0933	181	127.4	51.4	89.16	4/20/1968	81	5
238	8-28-68 Wed 0959	136	123.7	51.7	88.5	9/1/1968	64	4
496	6-26-72 Mon 1453	212	121	51.6	89.6	7/2/1972	96	6
573	6-15-73 Fri 0558	204.6	122	51.6	89.5	6/17/1973	32	2
613	11-30-73 Fri 0519	183	121	51.8	89.1	1/29/1974	953	60
638	4-3-74 Wed 0729	202	121	51.6	89.4	4/13/1974	158	10
656	5-27-74 Mon 0724	220	121	51.8	89.7	5/29/1974	32	2
672	8-12-74 Mon 0624	148.5	123	51.8	88.6	8/18/1974	97	6
772	9-29-75 Mon 0417	199	125	51.6	89.4	10/2/1975	48	3

Apogee, perigee in sm, inclination in degrees, period in minutes. Days may not be whole days. Launch vehicles generally assumed to be the standard Soyuz launcher but in the case of Cosmos 496 (and perhaps others) some data shows the use of the Zond (Proton) booster.

TABLE XI: THE SPACE STATIONS CALLED SALIUT—IN ORBIT FROM 1971 THROUGH 1975

Saliut No.	Launch Date GMT		Ha	Hp	Inc. (deg)	Period	Saliut Manned by Crew*
1	4-19-1971, Mon 0139	Initial Orbit	137.8	124.3	51.6	88.5	Soyuz 1; 23 days, June 7-29-1971
		"Final Orbit"	172	158	51.6	89.6	
2	4-3-1973, Tue 0900	Initial Orbit	161	133.6	51.6	89	Never manned; damaged in accident in orbit
		"Final Orbit"	184	162	51.6	89.8	
3	6-24-1974, Mon 2237	Initial Orbit	167.8	135.4	51.6	89.1	Soyuz 14: 15 days, July 5-19, 1974
		"Final Orbit"	172	158.4	51.6	89.7	
4	12-26-1974, Thu 0414	Initial Orbit	167.8	136	51.6	89.1	Soyuz 17: 29 days, Jan 12 to Feb 9, 1975
		"Final Orbit"	221.5	214	51.6	91.3	Soyuz 18: 62 days; May 26 to July 26, 1975

* Indications for manning are for time actually aboard the Station. Apogee, Perigee in sm, inclination in degrees, period in minutes

9

SATELLITES FOR THE PEOPLE

Despite the multitude of scientific and military uses of spacecraft there still remains a huge blank area in space activity that is yet to be filled by the endeavors of man. None of the space powers have fully applied their talents in using space for commercial and industrial purposes. There are many manufacturing processes that are possible in space and are not possible on the Earth's surface (or are only at enormous expense) that would lead to products useful in every aspect of civilian life.

Though transistors have replaced the vacuum tube in most electronic areas, still many power tubes have yet to find their analogue in the transistor. The transistor is currently limited to devices using very low amounts of power. This is not true for the vacuum tube. However, the latter "wears" out because of the residual amounts of air—among other items—that remain when its glass envelope is sealed. The better the vacuum within a tube, the longer its useful life. One might literally take vacuum tubes to orbit and break the glass to let the residual air out and then reseal the tube to improve its life. Of course, that's silly. Better to manufacture the tubes in Earth-bound plants and then take them to orbit for sealing in the endless vacuum of space.

Though it's recognized that an orbital manufacturing plant would have some atmosphere of its own surrounding the spacecraft, still it would be relatively simple to isolate a tube sealing device from that atmosphere. The vacuum in an orbitally sealed tube would be thousands of times improved in terms of the vacuum obtained within the tube. Its lifetime would increase in direct proportion to that harder vacuum. Once such an orbital plant is established, it can be shown that the costs of instruments using such orbital vacuum tubes would decrease considerably.

Additional possibilities for orbital manufacture include the making of highly improved ball bearings because of the readily available weightless environment, and, for the same reason, the production of materials with unusual properties, such as glass foam, which, once hardened, would be strong and highly insulating. Also foamed steel, once hardened would be a relatively light but very strong material. Thus, the latter could replace Earth-manufactured honeycomb materials that are both very expensive and not easily made. A given amount of steel in foamed form would replace shapes that would require expensive chemical milling on Earth; thus, the process would be made much less expensive and would conserve both the energy used in Earth-bound manufacture and large amounts of the steel itself because of the low density of foamed steel.

Growing large crystals for use in the manufacture of all members of the transistor family of devices could be accomplished in weightless conditions whereas the gravity immersed Earth's surface severely limits the size of crystal growth that can be accomplished in Earthbound factories.

Much of the precision in fine instruments is limited by the fact that gravity governs the accuracy of various shapes that can be machined on Earth. Such limits vanish in the weightless state on orbit. Sometime in the future, when space stations are really large enough to accommodate large numbers of persons and when shuttles to orbit can take people up with low accelerations, many medical studies will be performed on the circulatory system without the inhibitory actions of the loads placed on the heart and the remainder of the body by the gravity of the Earth's surface.

These last remarks represent only the barest scratching of the surface in what might be accomplished in orbit for the benefit of mankind. Some further extension of such thought can easily lead one to consider orbital resorts for the public, really large observatories for probing the farthest reaches of the universe, excursions around or to the Moon, manufacture of ultrapure vaccines, superfast delivery of important mail using orbital shuttles, and other related kinds of activity.

It is worth noting that many persons will be skeptical of the likelihood of these ideas ever coming to fruition. Many will never be convinced—no matter what logic and rationale or evidence is placed before them. To all such persons I would like to quote the eminent science fiction writer Arthur Clarke (who originally thought of the idea of communications satellites and the technique for using them); "Any sufficiently advanced technology is indistinguishable from magic."[76] This is particularly true for the unimaginative or unknowledgeable person but applies to many a well-established scientist as well.

It is worth quoting an example of the last statement. Harvard's [College Observatory] famed Astronomer William H. Pickering

---

76 *Profiles of the Future: An Inquiry into the Limits of the Possible*, Arthur C. Clarke, Bantam, 1962.

wrote, in 1908 (in reference to the Wright brothers airplane flight), "The popular mind often pictures gigantic flying machines speeding across the Atlantic carrying innumerable passengers in a way analogous to our modern steamships. It seems safe to say that such ideas are wholly visionary, and even if a machine could get across with one or two passengers, the expense would be prohibitive to any but the capitalist who could use his own yacht."[77] Well, it's damn safe to say that there are a lot of Pickerings still around.

The fact is that it is the unforeseeable that will, no doubt, bring to man his greatest advantages from space travel and the use of facilities in near Earth orbit. But how can one discuss the very things that elude one's imagination? Some time, some tomorrow, those answers will make themselves known. Today we live in ignorance of the future's wonders. But there are many uses of near-Earth space that are evident to us and that we are learning to use for our mutual advantages.

The two most obvious types of satellites that are of considerable aid to the public in the commercial sector are communications and weather satellites. Not only have the two major space powers made use of such satellites for some years now but also even the lesser space powers, such as France and Japan, have extensive plans for the use of these classes of satellites. The communications satellite has brought the use of overseas telephone conversations to a far wider spectrum of people simply because the price has been so substantially lowered. The old undersea cable price was of the order of $15 for 3 minutes between the US and Europe. The same conversation is now [in 1975] carried on for something under $6 via satellite.

For the Soviet Union, though costs are not available, their situation must have brought far better results. Few transcontinental telephone

---

77 *Enterprise, Jerry Grey, Wm. Morrow and Co., 1979*

lines for public commercial use exist across Siberia. The only other method previously available was the use of high frequency radio communication; a very imperfect technique, always full of static, subject to interference of many kinds, and relatively expensive. The introduction of the Molnyia (lightning) communications satellite and its commensurate ground station network (called Orbita) alleviated that problem to a very great extent. With the introduction of this system, the Soviets eased the problem still further by using a large payload—at least 2,000 lb and probably greater by the present stage of development, which permitted a greater output of broadcast power from the satellite.

While the US started with output power in the 3 to 6 watt range in their satellites, the Soviets started with output power of 40 watts. There is more to that statement than meets the eye. The smaller power used in the US necessitated utterly huge antennas on the ground to receive the signal from the satellite at a synchronous orbit altitude—22,000 miles. Large antennas cost large sums of money to build, operate, and maintain. As a consequence, the investment was huge and initial operating costs for the use of such telephone lines was not cheap.

The route chosen by the Soviet Union was influenced by several additional factors. To get to synchronous orbit their rockets would have to make a plane change of some 46° since that is the latitude of the Tyuratam launch complex. That would reduce the size of the payload that could be orbited and hence reduce the power output and, therefore, increase the needed size of ground antennas. The cost in propellant weight of making a plane change of just 1° for a rocket booster is considerable and, of course, such weight devoted to propellant must then reduce the payload that can be brought to orbit. The 46° penalty is huge. Thus, placing their satellites in a 12-hour orbit with apogee over the Soviet Union does several things for the Soviets:

1. Three such satellites in orbit spaced 120° apart ensures that at least one of them is always in view of the communications network.

2. The 12-hour orbit does not demand the propellant capability that the synchronous orbit satellite demands, hence the payload is considerably larger in the case of the Soviets—about four times as large as the same rocket could place in synchronous orbit. This latter fact neglects the extra propellant that would also be needed for the 46° plane change.

3. Because the payload is large and the commensurate power output is 40 watts, the ground antennas are relatively small. About 40 feet in diameter compared to the 85 foot diameter antenna used with the smaller US output power. Eventually the US situation can be expected to change with increased output power, but generally this is not the case at present.

4. The Soviet highway system is not extensive through Siberia but airports are relatively numerous. Every town of any substance has one. Thus, because the receiving antennas are not large, they are air transportable. They can then be manufactured at a sophisticated facility and flown to the most remote locations for assembly with the remainder of the station.

5. The smaller size of the antennas permits operation in very severe wind conditions with minimal protection, thus minimizing downtime in the event of bad weather. It should be recognized that times of bad weather are just those intervals when good communications are often not just necessary but critical.

In 1967 the Soviets inaugurated their Orbita system with an initial ground system network consisting of 20 stations, many located at distant and widely scattered Siberian cities. The network has grown considerably since that time; over 40 stations were scattered throughout the Soviet Union by the end of 1975. In 1969, the Soviets built such a station in Mongolia at Ulan Bator [Ulaanbaatar]—the first such station outside the Soviet Union; and probably only the beginning of a vast network outside the Soviet Union proper.

The Orbita communications system is used to transmit telephone conversations, telegraphic information, multichannel radio, newspaper facsimile photos, television and—mixed in with all this—military communications. A particularly notable use of the system is the transference of information, TV, and otherwise, between their ground control center and Soyuz spacecraft during their flights in Earth orbit.

In order to achieve a communications system with anything like the great reliability and versatility of the Orbita system using only a ground system, it would mean the laying of thousands of miles of transmission lines plus the repeater stations necessary for boosting power every so many miles. Substituting microwave stations without any intervening transmission lines would again mean thousands of stations (a thin system might get by on hundreds). Either one of these substitute systems would be enormously expensive to install initially. One must add to this the costs of maintenance—no little part of which would be due to the extreme weather conditions throughout Siberia much of the year. The difference in costs between the satellite system and any comparable ground system must be in the hundreds of millions of dollars at least, and a number in the billions would not surprise me at all.

The 12-hour Molnyia orbit is such that apogee at about 25,000 miles is located in the northern hemisphere while perigee at about 300 miles is located in the southern hemisphere. Assuming the first apogee (after placement in orbit) is over the Soviet Union, the second apogee will be located over North America. Molnyia orbits initially were inclined at 65° but are now [1975] at 62.8°.

The first operational *Molnyia 1* satellite was placed in orbit on April 23, 1965. However, a backward glance in time to August 22, 1964, at the launching of *Cosmos 41* is revelatory.

Cosmos 41 had an orbit that is completely described by the numbers defining the Molnyia satellite orbit. When it was placed in orbit, however, there was no announcement concerning a communications satellite. The numbers were puzzling, and try as I could, the mission of the satellite remained a mystery. None of my colleagues could discern the mission either. Our guesses were bad too. A failed lunar attempt, an interplanetary flub, a try for synchronous orbit were some of our wrong guesses. The timing was wrong for the lunar try as there was no interplanetary launch window at the time. The synchronous attempt could not be ruled out, although the orbit inclination, under normal considerations of the time, seemed intended—a synchronous attempt should have led to a much lower inclination; even with a major failure, the launch azimuth would have immediately resulted in a lower inclination.

One other observation of considerable interest was made at the time. There were several pieces of "debris" in the same orbit plane but with much lower altitudes; an apogee of about 280 miles and a perigee of about 125 miles. That, at least, was well understood. Whatever the mission, the total satellite was first brought up to a parking orbit, separated from the last stage of the booster rocket and then boosted to the higher orbit. The debris consisted of the last stage of the booster that brought the entire package to the parking orbit plus some interstage connecting hardware and the protective shroud for the communications satellite.

Insofar as the shroud is concerned, it is equally likely that the shroud was ejected while the satellite was on the way to orbit, which is a common procedure in satellite launchings. The shroud only serves to protect the payload from aerodynamic damage. Once out of the atmosphere—roughly above 200,000 feet—the shroud becomes excess baggage and the penalty for carrying it any farther is high. One can think of reasons for keeping a shroud but that is another story.

It was a fourth stage that took the satellite to the 25,000 mile apogee. It seems that this stage remains attached to the communications satellite after boosting it to high orbit, for no SPADATS bulletins indicated another object in a similar orbit. For the last few Molnyia satellites launched, there were secondary objects in essentially the same orbit, however, I believe that these are some sort of piggyback payloads, with missions separate from that of the communications satellite.[78]

All of the Molnyia satellites—there have been 50 through 1975— have been placed in orbit in the same manner as *Cosmos 41*. There were 31 announced as Molnyia 1 satellites. That left the implication that there were to be improvements in this series of satellites that might be called Molnyia 2, for instance. On November 24, 1971, the Soviets announced the launching of the first *Molnyia 2;* this was followed by a further upgraded satellite called *Molnyia 3,* first orbited on November 21, 1974.

---

78 There is some uncertainty about this since some information indicates that the last stage of the booster to the parking orbit is reignited to attain the 25,000 mile apogee. In that case, it may be only in recent launchings that this last stage was detached after final orbit was reached. That would account for the secondary objects in similar orbits as the main satellites. My personal view is to reject this explanation of the launch technique.

These 50 Molnyia satellites can be located in Table IX, listing all Soviet launchings through 1975. Because the Molnyia does have a perigee at 300 miles, its orbit changes slowly due to aerodynamic effects. As a consequence, the Soviets have placed a correction engine aboard the satellite. It has a thrust of 200 kg—440 lb—and a total working lifetime of 65 seconds. It should be noted that most corrections require 1 or 2 seconds of burning (working) so that the engine is adequate for many corrections. Such correction is likely required no more frequently than once every few weeks. This communications satellite has three transmitters, one operating and two in reserve. They each broadcast at 40 watts.

There are two 1 meter diameter parabolic antennas aboard the *Molnyia 1*, optical sensors for detecting the Earth, sensors for detecting the Sun, an electronic computer to receive and act on programs fed in from Earth stations, and a very sensitive receiver. The receiver has a backup unit. The entire system receives electrical power from a large star-shaped bank of solar arrays coupled with advanced chemical batteries such as nickel cadmium. Despite the proclaimed advantages of the 12-hour satellite the Soviets have commenced placing synchronous satellites in that orbit. *Cosmos 637* and *Molnyia 15* and *Statsionar 1* initiated this effort; the latter the first operational unit. These aid in around-the-world contact with space-tracking ships, military use, and pose some competition with Intelsat. Contact with the Soviet's enormous fish fleet is another evident use.

There are five other satellites to be considered in discussions of the Molnyia; e.g., *Cosmos 174* and *260*. Both have orbit parameters that are exactly like the Molnyia's but don't bear that name. While they might be failures, I believe that they were restricted for military use. The obvious question that arises is that it should make no difference what they are called independent of selected usage. Military communications can and are coded in any case. Specific frequencies are eventually detected by monitoring stations hence the question for their being termed Cosmos vehicles remains.

Up to the 12th *Molnyia 1*, all of the launchings were from Tyuratam. On February 19, 1970, the 13th *Molnyia 1* was launched from the Plesetsk launch complex and since then all but four have been launched from Plesetsk. *Molnyia 2* and *3* are also launched from Plesetsk. I attach no special meaning to this procedural change. Tyuratam activity apparently has been high and this operational system has been moved as a matter of convenience to make way for new research projects.

Weather Satellites

The efforts of the Soviet Union in the area of meteorological satellites are not nearly as clear as their communications satellite program and it is difficult to fathom the reasons for this. Both programs can be, and no doubt are, used in military operations; still, the beginnings of the weather satellite efforts are almost obscure when compared to the one other obvious applications satellite, the Molnyia.

Weather satellites, before they were identified as such, are traceable all the way back to *Cosmos 44*; but not in a manner so positive as to exclude the possibility that other experiments, military or otherwise, were being carried out. In general, the orbit parameters are quite distinguishable and the satellites are announced by the Soviets as being part of their meteorological system. Not always, however. Many satellites that from all the available evidence—it's pretty thin, consisting only of the orbit parameters—to be part of the system are not identified as such.

The current weather satellite system is called Meteor by the Soviet Union and their system has had 23 launchings identified by that name.[79] The Meteor satellites are launched into fairly circular orbits at an inclination of 81° and a fraction, at a period between 97 and 98 minutes, and at an altitude of approximately 400 miles. *Meteor 5* and *Meteor 10* were placed in higher orbits, being circular at 550 miles and, therefore, with longer periods; both close to 103 minutes.

Instrumentation is very complete permitting examination of cloud cover on the dark side of the Earth as well as on the Sunlit side something the US did not do till well after the Soviets. All the birds have been powered by a combination of solar arrays and chemical storage batteries; again, nickel-cadmium or silver-zinc as with the Molnyia satellite.

The first experimental weather satellites were not launched into 81° orbits but started at 65° orbits and at a variety of altitudes. *Cosmos 44* in August 1964 was placed in such a 65° orbit and in the high orbit that was typical of later weather satellites. Of course, at the time it was not easy to distinguish the mission but on the skimpy evidence of comparison with American weather satellites and some articles in the Soviet press, I concluded, correctly or not, that this was an early experimental weather bird.

The next such satellite was *Cosmos 58*. It had similar characteristics, although the average altitude was about 80 miles lower. *Cosmos 58* was launched at the end of February 1965. It wasn't until the end of that year that *Cosmos 100* was launched in mid-December 1965 and again the orbital characteristics of an experimental meteorological spacecraft appeared. Again, in mid-May 1966, another satellite, *Cosmos 118,* was placed in a highly circular orbit at just under 400 miles, and once more examination of the mission led to thoughts of a

---

79   Through 1975.

weather bird. This time the thoughts were enhanced by unconfirmed rumors that weather photos had been received from the *Cosmos 118* satellite. Since these rumors emanated from Moscow, I tended to believe them. If there was any truth in this, the Soviets should be ready to announce the successful operation, if not the actual launching, of the very next weather satellite that they placed in orbit.

The wait was shorter than I had anticipated. It seemed to me that nothing would occur until the end of the summer of 1966. Instead, *Cosmos 122*, launched on June 25 into a 65°, 388 mile circular orbit, with a 97 minute period was announced as a weather satellite. Since it was still under the Cosmos aegis, I could only assume that it was still very much an experimental affair as opposed to an operational system. Actually, the initial launch was made without identifying the satellite as a weather bird. It took two months before the Soviets made an announcement stating in *Pravda* of August 18, 1966, that *Cosmos 122* was a weather-investigating spacecraft. The announcement said in part:

> Taking account of the results of the check-out of the onboard systems, and the normal operations of the (previously) mentioned meteorological apparatus, it was recommended that experimental utilization of these measurements be started in the weather bureau. The World Meteorological Center in Moscow has been given word to include the most interesting results obtained in the mentioned measurements in the information intended for communications to organs of the weather service of the USSR and other nations.

Such weather data as mentioned in the communique was forwarded to the US National Environmental Satellite Service center in Suitland, Maryland, on August 18, the same day as the Soviet announcement. Although initial quality of the Soviet *Cosmos 122* photos was not too high, it did represent a beginning of real exchange of data in the weather satellite area and, as expected, it improved with time and the advance of equipment on later meteorological satellites (metsats).[80] The Soviet Union was now certain of its capability in this area to the extent that it released photos of the satellite itself, taken, of course, before launch during checkout at the launch complex.

A discussion of the instrumentation complex aboard the *Cosmos 122* covering not only the meteorologic instruments but also the controls, programming techniques, onboard digital computer, and the stabilization system was published in *Pravda* of September 24, 1966. The sophistication was very evident, particularly the control system for satellite stabilization, which included a method using nonpermanent magnets that interacted with the Earth's magnetic field—when they were "turned on"—and cold gas thrusters. The article was detailed enough so that one might have expected it to appear in a technical journal rather than an everyday newspaper.

All of these earlier metsats were launched from Tyuratam. A strong indication that the system had reached a desirable level of maturity came with the announcement of the launch of *Cosmos 144*; it was launched from the Plesetsk complex in to a new orbit of 81° inclination. The 65° metsat was now history. All metsats would now be placed in 81° orbits and all would be from Plesetsk. *Cosmos 144* was launched on February 28, 1967.

A few months later at the aerospace show at Le Bourget outside of Paris, the Soviets revealed many things; among them a full-scale replica of *Cosmos 144*. The workmanship was superb, the satellite an intricate combination of sophisticated instrumentation, and the

---

80 The lesser than desired quality was attributed to the transmission between countries and not the capability of the satellite system itself. The original satellite photos were said to be very good; the communications link between countries was far from good technically.

size quite impressive. I should judge from the photos published in *Aviation Week & Space Technology* that the satellite weighed close to 2,000 pounds. Its 2 meter diameter (6.5 feet) parabolic antenna assured an accurate reproduction of the orbital data at the Earth receiving stations. Star trackers, horizon sensors, optics up to 1 meter (about 3 feet) in diameter and advanced IR equipment were all the marks of the most up-to-date techniques for examining clouds in daytime and at night, the thermal character of the Earth from orbit, solar radiation reflected off of the clouds, and other related traits of any weather system. The system was more than a match for anything produced by the US at the time.

Now that the second Cosmos announced as a metsat had achieved the status of successful operation, the initiation of a fully independent weather satellite system with its own name, in the manner of the Molnyia communications satellite, seemed an appropriate prospect. That did not occur. *Cosmos 156*, launched in April 1967, was identified as a metsat still under the aegis of the Cosmos program. Some element of the system presumably was not operating in a fully satisfactory fashion. In due course this situation was changed.

On June 4, less than two months after the *Cosmos 156* launching, the "Main Administration of the Hydrometeorological Service of the USSR Council of Ministers" (reported just that way by TASS) announced the formation of "an experimental space meteorology system designated 'METEOR.'" The article describing the Meteor system (again, as in the *Cosmos 144* case) gave great detail on the equipment aboard *Cosmos 156*, noted that the system was formed by *Cosmos 144* and *156*, and went on to point out that the system was integrated into the national economy. The latter statement was emphasized in a short discussion of Soviet airline routes and the evident need for knowledge of weather conditions not only over the Soviet Union but also for the Soviet's international/intercontinental air traffic.

The satellites were also used to forecast wind and wave conditions for the initial stages in building a large floating dock and (apparently) moving it along the Soviet's Pacific coastline. The disposition of icebergs and other ice in northern sea routes was established using the metsats. It was evident that they were pleased with their new system and its wide use in several aspects of the Soviet economy, particularly as such uses could be appreciated by a broad spectrum of their citizens.

In a pleasing series of admissions, the article went on to detail the areas where problems had cropped up in earlier and rather embryonic metsats; stabilization, interaction with the Earth's magnetic field, aging of solar cells, lubrication in the hard vacuum of space, and the difficulties met in temperature control of the onboard electronics. The reader was also assured that all of the problems had been rendered amenable to satisfactory solutions. Thus, the creation of the Meteor system.

The two satellites were in orbits separated by about 95°, which permitted the determination of weather conditions for the same area by each satellite at six-hour intervals. Additional satellites, with altered spacing, would permit observation of conditions at smaller intervals, necessary for airline routes both domestic and otherwise. Despite the apparent success of the initial Meteor system the weather service was kept in an experimental mode through three more launchings; *Cosmos 184* on October 24, 1967; *Cosmos 206* on March 14, 1968; and *Cosmos 226* on June 12, 1968. These metsats were all determined to be experimental and I surmise that some fine tuning of the system was still in progress related not only to the satellites themselves but also to the handling of the enormous amount of data that is necessarily brought to the ground stations, to the interpretation of the data, and then to the dissemination of useful information derived from the weather system.

Finally, on March 26, 1969, the Soviets dropped the experimental status for their weather system by launching a metsat and calling it Meteor. This was the first of 23 Meteor satellites launched until the end of 1975.

By the end of 1969, the Moscow receiving center for the meteorological data was supplemented by another station at Novosibirsk in Siberia at about latitude 55° north and longitude 83° east. Additionally, it was noted also that Khabarovsk in the [Russian] Far East at latitude 48° north and longitude 135° east was being prepared as still another receiving center for the weather information. Both Siberian cities were to have very large receiving antennas installed—a fact of history by this writing—along with the necessary ground complex equipment. It is a safe assumption that other centers have since been added to the system. Data and the relay of such data have been simplified to the point that, for instance, every large Soviet airport regularly receives information on weather covering all of their air routes.

As a closing note on the weather satellites and the Meteor system, I would like to point out a situation analogous to one I observed when discussing the Molnyia communications satellites. Three Cosmos satellites—*Cosmos 389* launched on December 18, 1970; *Cosmos 405* launched on April 7, 1971; and *Cosmos 476* launched on March 1, 1972—all have the complete orbital characteristics of Meteor satellites but have not been so named.[81] The question arises here, as it did with *Cosmos 174* and *260* vis-a-vis the Molnyia, are

---

81 I have said on numerous occasions throughout this book that the Soviets frequently have surprises in store. Thus, another one appeared in the press of March 24, 1972, when an announcement stated that *Cosmos 476* was an ELINT satellite of "gigantic" size 10 to 12,000 lb. (See Chapter 7, "Soviet Reconnaissance Satellites: The Military World" on military satellites.) Since past ELINT birds were under 1,000 lb the adjective *gigantic* is accurate. Increased sophistication must be considerable. The same question can now be raised about *Cosmos 389* and *405* but still

these failures or are they dedicated to military use? I would choose the latter and add that the information that they obtain may well be much improved over that used in the Soviet "commercial" Meteor system. If they are military satellites then their use is evidently intended for the Soviet Navy and particularly, I'd guess, for their submarine fleet.

The dependence of the population of the peoples of this planet on agriculture make that field not only of prime importance but also to coin a phrase, a first among firsts. The vagaries of weather, insect plagues, forest fires, floods, frost, hurricanes, tornados, and improper or inadequate use of artificial fertilizers, all contribute to lessening both the quality and the quantity of the products of the planet's efforts in all aspects of agriculture. A forest valued for its lumber or a truck farm valued for its vegetables can be attacked by any of the aforementioned problems where substantial, and frequently total damage eventuates before the forester or farmer is aware of what is happening.

Efforts to limit these crop-damaging events through the use of methods confined to ground observations and subsequent actions have proved to be largely insufficient because they fail to examine and act on the particular problem in a context that recognizes the results of their actions on other crops and natural resources. The wide use of pesticides and the consequent pollution of adjacent rivers and streams in the run-off process is one clear example of the narrow attack on some agricultural problems.

A very good typical example of what orbital sensing can do to aid the agricultural community is given by the following: An aerial photo survey, not too long ago, of some Brazilian coffee plantations clearly revealed some areas infected with coffee rust disease, which, in the

---

remains unanswered; the Soviets are endowed with a capability for the unanticipated.

developed photographs, showed up as white against the healthy red trees. Processing and interpretation of the photographs took only a few days. The social and economic importance of this can be realized quickly when one considers that 27 percent of Brazil's foreign earnings come from coffee sales.[82]

The same amount of time spent by an airplane in this aerial survey if allotted to survey by an Earth resources satellite would have permitted surveillance of all of the Brazilian coffee plantations thus assuring that the plant disease was confined to just the small area noted by airplane or informing the authorities of a much more serious infection that would then receive immediate attention.

The search for new oil and mineral deposits; more important, the search for new supplies of water—especially far from seacoasts—is a tedious and expensive and, more often than not, unrewarding series of tasks.

Recognition of the direction of urban sprawl in conjunction with the surrounding land is poorly done at best when attempted using ground surveying and other presumably coordinated ground observation techniques. This is emphasized in so-called modern building methods where very large subdivisions, airports, and other large land consuming projects are initiated, and in fact completed, without any attention being given to effects on surrounding territory.

The laying of long-distance pipe lines for oil, the erection of cross-country transmission lines for electrical power, and the building of large dams are projects almost notorious for their neglect of the land that they are supposed to fit into or traverse. Dams have often been constructed—and in many places still are—everywhere, without one iota of consideration of the effects on life, both human and animal,

---

82 *Spaceflight* vol. 14, no. 2, February 1972.

in the rush to form a lake for hydroelectric power or to form a reservoir of water for near or distant populations.

The popular term "big picture" is as descriptive as any in trying to say what it is that all of these large projects fail to take essential cognizance of while under the unfortunate misapprehension that they are doing something to improve life for some community.

Each of these concerns, from those in agriculture to those in dam construction, can be resolved to a very large extent if a wider viewpoint was applied to each of the problems. And this can and is being done. The US is making a substantial investment in the area of unmanned satellites so that these problems can be examined from orbital altitudes using instruments designed for the purpose (two of these have been placed in orbit for such task resolution) and the Soviet Union, with relatively little fanfare on the topic, has initiated such studies using the Soyuz spacecraft and the cosmonauts aboard them.

TASS announced, on October 13, 1969, that the crew of *Soyuz 7* had made geological-geographical surveys, including photography of "characteristic sections" of the Earth's surface. These observations covered borders of the snow cover (in the northern regions), cloud cover (in relation to weather over various parts of the Earth—the latter no doubt referring to the Soviet Union), and recorded measurements of Sunlight reaching the Earth's surface. Assessments of cyclone formations and their progress over time were made in large cloud formations.

Photography of the Caspian Sea area permitted determination of conditions existing in the water and adjacent land areas. Studies were made of parts of the Soviet Union to ascertain the existence of ore deposits for future exploitation. These areas were not identified nor was the ore or ores that they searched for—or found. Only the remark that "these areas were relatively difficult to reach by other means," was made, which is almost no clue at all to identification.

All of the information released was too general to permit me to note precisely the experiments being carried out—typical of many Soviet announcements—but was not so obscure that one could miss the fact that they were, in fact, up to their ears in Earth resources experiments.

Some of the releases made in conjunction with the triple flight were really way out: Soviet specialist on aerial photography Nadezhda Lavrova talked of using the results of this kind of space research to eventually change the climate of the planet presumably to favor the Siberian regions of the Soviet Union. [Atmospheric Physicist] Kirill Kondratyev of the Academy of Sciences of the Soviet Union discussed the use of these spacecraft to study (warn of ?) floods and dust storms to forecast high water by observing the differences between snow and ice caps, to locate the best fishing areas, and to ascertain soil conditions in order to determine optimum planting and harvesting times.

Though it is not an Earth resources experiment directly, the *Soyuz 6* crew made meteorological observations that aided directly in the landing of that spacecraft at the end of its flight. Hurricane Jennifer, off the West Coast of Mexico [in 1969], and Hurricane Inga, off the Bermuda coast [in 1969], were traced and reported on during the flight of the three Soyuz spacecraft. So too were several tropical storms in both the Atlantic and Indian Oceans.

In order to make observations on sea conditions (and fishing conditions?) and check them against those made on seaborne stations on the various oceans, the troika were in contact with seven widely spread Soviet research ships—the *Morzhovets*, the *Nevel*, the *Bezhitsa*, the *Dolinsk*, the *Ristna*, the *Kegostrov*, and the *Borovichi*.

Contact was made with the ships and with Moscow through the use of the Molnyia network, certainly a complex and sophisticated system implying great capabilities.

Similar experiments were carried out aboard the *Soyuz 9* flight but were not dealt with at any length in the press. Those Earth resources experiments that were carried out aboard the *Soyuz 11/Salyut* (and there must have been many) never were discussed in detail because of the tragic ending of that mission, which completely obscured all other events during that record-breaking and otherwise successful flight.

A review of information given out to the public on the results of the Soyuz-conducted Earth resources experiments makes it quite clear that a wide program was carried out—and will certainly be continued in future flights—but that specific details are not going to be made available for a long time, if ever.

It doesn't take much thought to see the connection between much of that kind of information and various military interpretations of the same basic data—hence the secrecy.

10

ZOND SPACECRAFT TO THE MOON

After some of their early failures in space missions, the Soviets adopted some general names to cover a variety of missions so as to obscure their purposes from the public—at least until some reasonable degree of success was attained in a particular mission. The most well-known of these cover names is the Cosmos series. The only other general name in use is the Zond series of spacecraft.

Zond—which means, appropriately enough, probe—was used to disguise, from the unknowledgeable public, some of the early flights to the planets Venus and Mars; *Zond 1* to Venus, *Zond 2* to Mars, and *Zond 3* as a diagnostic payload to test spacecraft operation at planetary distances. As I noted earlier in Chapter 4, "To the Planets," *Zond 3* worked for a long time although by its very mission it was not intended to approach any particular planet.

For a long time, the spacecraft *Cosmos 21* was a considerable puzzle, for no planetary launch window was open at its launching; other aspects of its mission—its orbital parameters and so on—didn't give any clue either. It wasn't until some years later that comparison of its launch time and the phase of the Moon at its launch time with those values for *Zond 5, 6, 7,* and *8* made it clear to me that *Cosmos 21* was the first intended lunar Zond. All the lunar Zond spacecraft have photographed various parts of the farside of the Moon; as a consequence, they had to be launched so as to arrive when the farside was lighted by the Sun in that area to be photographed. *Cosmos 21*, launched on November 11, 1963, and decayed on November 14, became part of the debris of lunar spacecraft history.

The first recognizable lunar Zond spacecraft did not show up for some years later and, as it turned out, it was literally a "practice" spacecraft—termed by members of the astronautics community a precursor. It was launched at 1828 GMT on March 2, 1968, more than four years after *Cosmos 21*. It too baffled me for some time, for the launching took place from Tyuratam, at a time when the relationship of Earth and Moon was quite different from that existing during prior lunar probe launchings. Prior lunar launchings commencing with *Luna 4*—the first intended soft lander—took place during the first-quarter phase of the Moon. The *Zond 4* launch occurred seven days before the first quarter.

The TASS report on the launching, in *Pravda* of March 3, 1968, did not mention the Moon or any of the other planets. The report simply stated that "The purpose...is to study remote areas of near-Earth space." Instead of reentering the atmosphere for an intended recovery, it had a major guidance failure and ended up in a highly eccentric orbit around Earth. It was still there at the end of 1975—an orbiting derelict, completely forgotten, out of the public eye entirely.

When the parking orbit debris from the *Zond 4* spaceshot decayed and was incinerated in the Earth's atmosphere, the burning debris (the last stage of the rocket that placed the probe in Earth orbit initially, some interstage structure and possibly some shroud material) passed over Kentucky, southeast Ohio, and diagonally over Pennsylvania from southwest to northeast. Immediately, the US Air Force and the press in those states received "news" of a spate of visits of unidentified flying objects (UFOs). The persons sighting the presumed UFOs were greatly disappointed when they learned that what they had seen was the debris of a Soviet space launching. Actually some 70 persons had reported the sighting of the flaming debris, and some who were airline pilots correctly ascertained that they had observed a reentry of satellite debris.

Zond 5, the first circumlunar probe, was launched on September 14, 1968, at 2142 GMT from the Tyuratam launch base. At 67 minutes after launching, *Zond 5* was injected onto a translunar trajectory. A midcourse correction occurred on September 17, 1969, at 0311 GMT when *Zond 5* was about 202,000 sm from the Earth. On September 18 the probe passed around the Moon, its point of closest approach was 1212 sm (1950 km). When *Zond 5* was 88,660 sm from the Earth a second midcourse correction was made, which altered the probe's velocity by 0.005 percent; the total velocity change was 1.15 fps (0.35 meters/sec).[83]

Entry into the Earth's atmosphere took place at 1554 GMT September 21 after separating the descent vehicle from the instrument compartment, at 1530 GMT. The spacecraft's parachute system deployed at about 1558 GNT and *Zond 5* splashed down at 1608 GMT in the Indian Ocean at latitude 32°38' south, longitude 65°33' east. The probe was lifted aboard ship on September 22 (Moscow time). The recovery ship that retrieved the probe from the sea was the *Vasily Golovnin*.

---

83 During the return trip.

There was a large group of ships that participated in the retrieval. In particular the research vessel (i.e., tracking ship) *Borovichi* tracked and located the *Zond 5* spacecraft. Splashdown occurred on September 21 at 1908 Moscow time. The best speed the *Borovichi* could probably make was 15 knots, but it was deterred by a force 4-5 wind that evening so that it almost certainly made no more than 10 knots and quite probably less. Since probe retrieval was made on September 22, it become evident that the probe landed approximately 50 nautical miles from the *Borovichi*.

Although the *Zond 5* carried a payload of photographic equipment, radiation sensors, and a spectrum of biological specimens, it seems rather extraordinary that a group of 85 Soviet scientists flew to Bombay [renamed Mumbai in 1995] to examine the probe and its payload before transshipment to Moscow. The scientists accompanied the probe and payload on the same [Antonov] An-12 aircraft.

Dr. Charles Sheldon [Chief of the Science Policy Research Division of the Congressional Research Service] of the Library of Congress, during a lecture given at Stanford University on September 26, 1968, stated that the *Zond 5* booster had a capability of placing 50,000 lb into Earth orbit. He also stated that this booster was an improved version of the "Proton" booster (the Proton booster had previously launched three high energy cosmic ray experiments into Earth orbit; each of these payloads weighed some 27,000 lb). The booster improvement therefore was apparently the addition of another stage.

Since single stage-to-orbit is an unlikely sophistication in Soviet technology and four stages for large boosters seem an off-optimum choice, one can assume that the Zond booster included three stages to Earth orbit. The Soviet Union in its discussions of the *Zond 5* and 6 flights clearly indicated that a fourth stage was used to inject the Zond probes into translunar trajectories. Consequently, I interpret

Sheldon's remarks, and it appears to be the logical result, to clearly infer that the improved Proton booster placed the *Zond 5* spacecraft plus the complete lunar injection stage into Earth orbit and that the total weight thus placed was equal to (or perhaps, slightly less than) 50,000 lb. Sheldon's remarks served to confirm my own calculations on the Proton (or as I prefer to call it the Zond) booster. Note my remarks in Chapter 5, "Earth Orbit Experiments."

Because the Soviets were so pleased with the results of the *Zond 5* circumlunar mission they published, in their daily newspaper, no less, a technical article on the mission and some of its results. I've included the article here to indicate the exposure that the Soviet public gets to scientific advances made in the Soviet Union. I took the liberty of excluding from the article some remarks on early history in biological studies together with some highly technical wording of interest to specialists in biology only.

"Biological Investigations in Outer Space (Zond-5)"

O. Gazenko, P. Saksonov, V. Antipov, & G. Parfenov

Pravda, no. 320, November 15, 1968, p. 3

As has already been reported, the "Zond-5" flew around the moon and splashed down in the Indian Ocean on 21 September 1968. Among the numerous scientific apparatus on the probe were containers with biological objects.

Within the total program of research work to <u>study and master universal space</u>, an obvious place belongs to the new area of natural science, space biology. Among its fundamental problems is the study of the biological foundations and principles to assure space flights, the investigation of conditions under which extraterrestrial life could exist, and also of peculiarities in its form...

Starting with 1960 (the successful launching of the second ship-satellite with animals and other objects on board) and up to the present, Soviet biologists have conducted researches on fourteen recoverable spacecraft. The "Zond-5" became the fifteenth on board which were live organisms.

Prior to the discovery of the Earth's radiation belts, ionizing radiation in outer space was not considered as a factor which could essentially affect flight safety. Since the discovery of the radiation belts, and particularly, the radiation accompanying chromospheric flares on the Sun, space radiation has become one of the principal barriers on the road to man's penetration into space. Therefore, obtaining data on the biological efficiency of various radiation sources in outer space is of great scientific and practical interest.

Voyages of spacecrafts with crews on board, <u>which would last several years</u>, will evidently become practically realizable only after partially, or completely, closed, reliably functioning, ecological systems have been created in which organisms of different complexity would be interconnected. Therefore, it is already now necessary to know how weightlessness and cosmic radiation, separately and in conjunction with other flight factors, will affect fundamental biological processes: the propagation, growth, and development of organisms; in what direction they can alter hereditary properties, the physiology of subcellular and cellular structures, and to what extent this is reflected in the stability of closed life systems.

At present the investigation of the biological objects to be utilized in flight experiments is the evolutionary, comparative physiology principles which would permit the most thorough estimation of the biological effectiveness of flight factors. Mammals (dogs, guinea pigs, rabbits and mice), reptiles (turtles), animal and human cells in culture, Drosophila flies, seeds of higher plants, spiderwort microspores, chlorella seaweed, numerous microbiological, and cytogenetic objects were utilized in the tests on ship-satellites (during 1960-1968).

It has been established as a result of the research conducted, that disturbances of a small, but statistically confident magnitude occur in the hereditary structures of diverse biological objects, mice bone-marrow cells, higher plant seeds, lysogenic bacteria, Tradescantia microspores et al., under the effect of various flight factors including cosmic radiation and weightlessness.

The experiment conducted in the "Zond-5" is a new step in realizing the program of biological, and particularly, of radio biological research in space. The flight of a craft through the radiation belts and beyond the limits of the Earth's magnetic field is of essential interest, for example. The following living objects were on the "Zond-5," turtles, Drosophila, meal worms, Tradescantia plants with buds, Kell [blood] cells in culture, seeds of various higher plants such as wheat, pines, barley, chlorella in different nutrients, lysogenic bacteria of different kinds, etc.

The physical parameters, the total dose of cosmic radiation, were investigated, and recorded with the aid of various kinds of dosimeters (direct reading dosimeters, thermally luminescent glass, photo dosimeters, nuclear emulsions, etc.).

Therefore, complete continuity has been retained in this experiment, relative to the objects and methodology of the experiments utilized in the flight tests conducted earlier....

In particular, diverse pharmaco-chemical anti-radiation means (cystamines, AET [adjuvant endocrine therapy], etc.) were used more widely and in a new aspect (as compared with past tests) to analyze mechanisms of the radio biological effects. Containers of new constructions for the prolonged residency of such objects as Drosophila, Tradescantia plants, lysogenic bacteria, meal worms under flight conditions were also successfully tested.

The first, and naturally, most difficult step in this unique experiment of flying around the moon and returning to Earth has been completed favorably. The biological objects have reached the laboratory in good condition, and a large collective of scientists has proceeded to process the quite valuable material with great interest. The analysis has only started, and now only some preliminary results can be announced.

Thus, the total dose of cosmic radiation recorded by utilizing dosimeters located at points where the biological containers were affixed, was several rad, which corresponds to nominal data. Despite the fact that the dose is small, the estimation of its biological (genetic) effectiveness is of definite interest to radio-biologists since, as we have mentioned above, it has been obtained primarily during the organisms' passage through the Earth's radiation belts.

After having been recovered on Earth, the turtles were active, moved about a great deal, ate with appetite. They lost approximately 10% of their weight during the experiment. An investigation of certain blood indicators... did not evidence any essential differences between the test control animals. At the same time, an elevated content of glycogen and iron in the liver tissues was detected in the experimental turtles in a histochemical analysis of a number of organs and tissues, made in the first twelve days after splashdown. The definite influence of the complex of flight factors was also exerted in the structure of the animal's spleens.

Preliminary analysis of the lysogenic bacterial cultures shows that the flight on the probe exerted a definite inductive influence on the phage production of microbes. There is a foundation to hope that final processing of the material would permit obtaining new, quite interesting, data on the biological (genetic) efficiency of near-Earth, and near-moon space factors. The biological experiment of the "Zond-5" prepared and conducted by the efforts of many Soviet biologists and physicists, produces an essential contribution to the further development of the fundamental aspects of space biology.

* * *

The flight of the Soviet circumlunar probe *Zond 6* was initiated on November 10, 1968, at 1911 GMT from the Tyuratam launch base. Its parking orbit parameters were apogee, 130.5 sm; perigee, 115 sm; inclination, 51.4°; period 88.31 minutes (all values were given by *Pravda* except the period, which I derived using the average altitude of 122.75 sm). *Zond 6* was injected into a translunar trajectory 67 minutes after liftoff at 2018 GMT. It is certainly logical to assume that the improved Proton booster launched *Zond 6* as it did *Zond 5*. A midcourse correction was made on November 12 at 0541 GMT when the probe was 152,500 sm from Earth. *Zond 6* circumnavigated the Moon on November 14; its closest approach to the Moon was 1504 sm.

I have calculated that a 14,000 lb payload can be injected into a translunar trajectory (specific impulse and thrust are given by G. Petrovich in an article titled "Home of Cosmonautics," published in *Vestnik Akademii Nauk Kazakbokoy*, no. 10, 1967, pgs. 54-62). The weights for this injection stage are propellant weight, 30,200 lb; weight of structure including tanks, rocket engine, interstage, 4,500 lb; weight of propulsion module controls, guidance system, 1,000 lb; and the weight of the translunar payload was then 14,000 lb. The total of these weights adds up to 49,700 lb. The mass fraction for this stage is 0.846; a value lower than might be expected for a comparable stage for a similar US circumlunar mission. This value is just the sort of thing that one might expect for the Soviet space vehicles; from a history of observations, they tend to use heavier structures and lower energy propellants than does the United States in its programs.[84]

Zond 6 made a second midcourse correction at 0640 GMT on November 16 after the lunar flyby while at a distance of 146,320 sm from Earth. During the return trip a third midcourse correction was made at 0536 GMT on November 17. Initial reentry into the Earth's atmosphere took place at 1358 GMT on November 17. Vehicle velocity was reduced from 36,736 fps to 24,930 fps. If the period from the initial entry to landing is taken as 30 minutes then the entire circumlunar trip from liftoff to soft landing in the Soviet

---

84 The mass fraction for a stage is the weight of stage propellant divided by the weight of the entire stage including propellant but not including the payload carried by the stage.

Union took about 163 hours 30 minutes. The probe probably landed south of the city of Kostanay, which was the Soviet announced landing area for circumlunar probe *Zond 7*.

The city of Kostanay is almost directly north of the launch base at Tyuratam. There was an announcement that biological specimens were carried on *Zond 6* but experiment results were not readily available in the literature until 1971. They were in essence a repeat of the *Zond 5* mission insofar as biological studies were concerned.

Meteor particle flux and energies were studied with great interest since *Zond 6* passed through the Leonids meteoroid showers during its flight. The *New York Times* of November 26, 1968, quoting TASS, the Soviet press agency, stated that *Zond 6* carried 93 feet of 7.5 inch-wide film in its cameras.

A *Pravda* article of November 24, 1968, stated that the flights of *Zond 4*, *5*, and *6* were also checkout flights for manned circumlunar flight.

Evidently, even at that late date, the Soviets must have thought that they had a good chance of conducting a manned circumlunar flight before the US could do so.

The *Zond 7* spacecraft was launched from the Tyuratam cosmodrome on Thursday, August 7, 1969, 2348 GMT. It performed a circumlunar mission practically identical to that of *Zond 6*. The spacecraft returned to Earth and landed north of the city of Kostanay on August 14. Kostanay is 700 miles north of Tyuratam.

The time of landing on the Earth was not announced but a comparison with the flight duration of *Zond 6* leads to an estimate for landing at 1900 GMT on August 14 yielding a total time of flight of about 163 hours. The use of touchdown rockets at the moment of soft landing was announced (UPI, Moscow, August 15, 1969, published in the *Palo Alto Times*).

Sir Bernard Lovell, Director of Jodrell Bank, reported that *Zond 7* passed behind the Moon at 1611 GMT on August 11 and reappeared at 1639 GMT.

Zond 7 used Earth, Sun, and stellar sensors for the purpose of navigation and orientation. It seems evident from the (minimal) information released on the *Zond 7* flight that it had an improved flight control system aboard with greater capability than that carried aboard *Zond 6*.

Although commentaries were made indicating that photography, space environment measurements, and radiation biology experiments were carried out, very few specifics were revealed by the Soviet Union until the 1971 articles referred to above. Again, the data seemed to be pretty much a repetition of the earlier experiments.

A few photographs taken by cameras aboard *Zond 7* were released; a rather spectacular photo of a full Earth rising above the lurain (lunar surface) has been reproduced in the September 1, 1969, issue of *Aviation Week & Space Technology*.

There was, however, a vast difference in the flight regimes of *Zond 6* and *7* from that of *Zond 5*. I never could determine whether or not *Zond 5* had been intended to return to and land in the Soviet Union directly as did the two later *Zond 6* and *7*. If *Zond 5* was intended to make a lifting reentry into the Earth's atmosphere, rather than the ballistic reentry that it did make, then one could say that the mission was not entirely successful. However, the Soviet proclivity for taking one step at a time might easily account for the ballistic landing for the first flight and the lifting reentry of the next two flights.

All three flights came back from the Moon in a path that carried them over the South Pole and then over the Indian Ocean. Such a path requires a change in velocity of over 10,000 fps in order to alter the flight path to bring it over the Soviet Union without a prior touchdown in the Indian Ocean. The technique demands an exquisite sophistication in guidance capability. The spacecraft must, at a precisely predetermined angle, enter the atmosphere, fly in a programmed manner to lose enough velocity to bring it down to a velocity just under that required for orbiting the Earth as a satellite, and then exit the atmosphere and fly a specified distance outside the atmosphere at the almost satellite velocity. If all has been programmed properly and carried out with precision then, because the spacecraft is not at full satellite velocity, it will reenter the atmosphere again and descend to the Earth at a prescribed location. A very few feet-per-second error and the spacecraft will land far from the desired location. If the error is still larger, then the spacecraft becomes a satellite and its landing becomes a question of days or even weeks—in a still worst case the satellite portion could be months. If the initial dip into the atmosphere is too steep the spacecraft may be destroyed completely. If the exit after the initial entry is incorrect then the spacecraft would fly into a huge elliptical orbit, which could make final retrieval a hopeless task.

The *Zond 5* atmospheric entry was described by the Soviets as encountering about 16 g's (16 times the acceleration of gravity) but nowhere did they state that the animal life aboard was harmed. On the contrary, the animal life was later displayed and described as none the worse for the particular reentry. A fair-size net of research-tracking ships was located in the Indian Ocean for the *Zond 5* return indicating therein a "reasonable" expectancy of an ocean landing. Of course, there could have been a land network also deployed expecting a landing in the Soviet Union proper. This latter force, if it existed, was never discussed with respect to *Zond 5*. Such a land network, including aircraft, was discussed for the *Zond 6* and 7 flights. The latter flights, because of their lifting reentry experienced a force corresponding to much less than half the g forces of the *Zond 5* spacecraft.

From the *Pravda* remarks in November 1968, it seems quite apparent that a manned circumlunar flight was in the offing at that time. Did the American plan to orbit the Moon 10 times in December 1968 cause the Soviets to cancel their proposed manned circumlunar flight or did a technical hitch appear to cancel this endeavor as announced in *Pravda*? This may never be known. If their ego was abraded by the thought of their circumlunar flight versus the American 10 orbit plan then I think that they made a considerable error. They should have flown and ignored the inevitable comparisons that the world would have made. A circumlunar flight at the time would have been admirable in any case. Alas, that they didn't try. Still, as I've said before in this book, the Soviets will land on the Moon. They will establish a base there before the Americans will. The one objection I must make with the movie, *2001: A Space Odyssey*, is that they had the Soviet and American roles reversed in terms of progress to be made in future spaceflight activities.

A little over a year later on October 20, 1970, the Soviets launched *Zond 8* at 1955 GMT. It looked very much like a repeat of the earlier Zond flights. It passed the Moon somewhat closer, at a distance of 696 sm and returned to Earth 162 hours later at 1355 GMT on October 27. However, there was an unexpected variation in the flight path—for the spacecraft returned over the northern hemisphere and landed in the Indian Ocean at latitude 13° south and longitude 73° east, about 453 miles southeast of the Chagos Archipelago. The Soviets announced that this path was a deliberate choice to allow flexibility for contingencies in future flights. It seems quite evident that the flight path in reentry was distinctly a ballistic one.

Again, turtles and other lifeforms were carried aboard. About a year later, [Physiologist] Oleg Gazenko, of the Russian Academy of Sciences, announced that the Zond flight program was at an end and the studies resulting from it were essentially complete—with satisfactory results. Full results and determinations were to be published "at a later date." Besides the collection of data and additions to knowledge from the Zond program, it seems apparent to me that all of the information from the program are suitable inputs into a manned flight program for some future date—not too far off, I hope.

Though the Soviets have presented papers on the results of experiments derived from their Zond lunar spaceflights, detailed presentations on or publications of their Zond spacecraft design are still completely lacking. In contradiction to this omission, they have displayed full-scale, finely detailed replicas of other spacecraft whenever the opportunity arises following a particular success in a pertinent sector of spaceflight. There is a notable gap in these displays—that of the lunar Zond spacecraft. These have never been publicly displayed leading me to the conclusion that the eventual intended use of the lunar Zond class of spacecraft has yet to be displayed in spaceflight missions.

Occasionally new evidence leads to identification of previously unresolved spaceflights. Such evidence may consist of entirely new material or is revealed by reviewing past material in a new light thus yielding new, useful, and revelatory correlations. In reviewing the lunar Zond flights, I reobserved that *Zond 5* to *8* all were launched during the same phase of the Moon (last quarter), at a time of maximum declination, and at maximum radial distance of the Moon from the Earth. *Zond 4*, however (never intended for a circumlunar flight), did not have similar parameters. It was launched with the Moon at 0° declination, at the phase of new Moon, but with the Moon not at its maximum radial distance.

Searching through "unsolved" launchings, I found that both *Cosmos 146* and *154* were launched at negative declination angles, at new Moon, and with the Moon at maximum radial distance, thus fitting into the Zond class if *Zond 4* and *Zond 5* to *8* are all included. Moreover, their launchings were separated by just 29 days, the spread of the month-to-month launch window to the Moon. I have assumed that *Cosmos 146* and *154* were precursor Earth orbit tests for the Zond series of spacecraft. The delay of one year between the launch of *Cosmos 146* and *Zond 4* gives rise to several possibilities:

- *Cosmos 146* and *154* were intended only to check out the lunar payload from Earth orbit—possibly at lunar return reentry speeds, or,

- Their Earth orbit tests, separate from propulsion activity, were unsatisfactory to some extent, or,

- Translunar injection did not occur because of an engine malfunction (a common event in Soviet spaceflight).

In considering the third item it is known that both spacecraft had engine starts in orbit; *Cosmos 146* on its 17th orbit. Because of the interval separating *Cosmos 146* and *154* from the appearance of *Zond 4*, it seems fairly obvious that the Earth orbit tests, whatever they were, did not satisfy the Soviets at the time. Nevertheless, the long burning times of these spacecraft at engine restart lead one to believe that they were connected with reentry studies and tests.

11

RETURN TO THE MOON—A BIG STEP

Each time a long gap appears in any branch of the Soviet space program, I am inclined to think that some unrevealed failures have occurred or, recognizing that programs—and their spacecraft—grow in size. I begin to wonder about what the next generation of spacecraft in the particular program will look like, how big it will be, and how its sophistication will increase.

Occasionally, a space program is completed, terminated for reasons unclear, or takes so divergent a path that making a connection to its predecessors is utterly obscured. That too blurs the picture. *Luna 14* had been launched on April 7, 1968. It was now over a year since that launching and long before the year-long interval had passed all these puzzles wandered in and out of my neuron complex. Competition for the manned Moon landing seemed uncertain, at best, for many events that should lead to such a landing had not occurred. Rumors of a manned circumlunar trip by the Soviets were plentiful but without hard evidence; this led to endless speculation on what might have occurred in various aspects of the Soviet lunar program.

I mulled over all sorts of thoughts but would not surrender to the frequently expressed nonsense in the press that some great catastrophe had taken place to bring this superb reach into the unknown to a halt. Though not really sure myself—but integrating all of the Soviet space program history that I knew—I came to the decision that *Luna 14* was the last of its kind and a new generation of spacecraft plus, of course, a substantial step up in capability, was soon to be clearly displayed. It's an odd feeling not being in control of something that you have a great interest but are literally hopping up and down on one leg waiting for the next event to come to the stage.

The launching of *Luna 15* at 0255 GMT on July 13, 1969, caused about as big a stir as one could imagine for it was just three days before the launch of the first manned lunar lander, *Apollo 11*. The competition was obvious and exciting to everyone, everywhere. The flight time was immediately announced by the Soviets in a TASS release and the announcement contained the statement "to make further scientific investigations of the Moon and near-Moon space."

That set of remarks gave rise to speculation that the launch used the booster called Webb's Giant, or the Zond booster, or the Soyuz booster—these each leading to payloads of vastly different sizes.[85] Almost a year after the event, Howard Benedict, an aerospace writer for AP, stated in an article that appeared in the *San Jose Mercury* of May 7, 1970, that the *Luna 15* launch was preceded by a launch

---

85 NASA Administrator James E. Webb had frequently forecast the launching of a Soviet booster far larger than Saturn V, having a thrust in excess of 10,000,000 lb.

using the Proton (i.e., Zond) booster that had failed, and that *Luna 15* was a substitute smaller shot. Benedict and his informants were entirely wrong.

Luna 15 was launched by the Zond booster and was larger than prior Luna spacecraft by a factor of approximately four; that is, it was about 14,500 lb in translunar configuration. During its flight from translunar injection to lunar orbit injection, which took 101 hours and 48 minutes, there were 28 communications sessions with the spacecraft. It was placed in lunar orbit at 1000 GMT on July 17 and so established the longest flight time to the Moon for any spacecraft up to that time. That was an obvious clue that the Soviets were trying to maximize payload by using the lowest essential velocity to get to the Moon. The first orbit was 126 by 35 sm at an inclination angle of 127°, and a period of 120.5 minutes.

Astronaut Frank Borman had just returned from a very friendly trip to the Soviet Union and used this nicely established astronaut-cosmonaut link to call M. V. Keldysh, head of the Academy of Sciences of the Soviet Union, to express the American concern of possible interference with the *Apollo 11* flight. Keldysh by return cable assured the NASA people that no such action was intended, that they (NASA) would be informed of all *Luna 15* orbital maneuvers without delay, and that the paths selected for *Luna 15* were well away from the orbital track of *Apollo 11*. He was true to his word: the Manned Spaceflight Center at Houston [renamed Johnson Space Center in 1973] was informed that new orbital parameters for *Luna 15*, made on July 19 at 1308 GMT, were 137.3 by 59 sm, at an inclination of 126°, and with a period of 123.5 minutes, and that a second change was made at 1416 GMT on July 20, leading to values of 68.35 by 9.94 sm, at 127°, and with a period of 114 minutes.

It was evident from the last set of numbers that a landing was imminent, and so it was. On July 21 at 1547 GMT *Luna 15*'s retro-engines were switched on preparatory to a soft landing. The landing at 1551 GMT, just 4 minutes later, was not soft. Exactly what happened was not revealed, leaving me to speculate on the series of events. Since the engine did come on I'll suggest that no propulsion failure occurred. Having eliminated that possibility, two others that are likely remain; a guidance failure so that proper control was lost and the spacecraft tumbled (even slight tumbling would spoil the landing) or a bad piece of luck wherein the vehicle landed on a steep slope and then tumbled down the slope and was lost. Of course, there are other possibilities and one could make an endless list.

During its 52 orbits around the Moon there were an additional 58 communications sessions with *Luna 15*. It was also pointed out—via TASS—that this spacecraft was different from previous Soviet lunar landers in (among other things) its ability to make changes in its orbit parameters.

The mission had ended according to the Soviet Union but it had hardly begun if one were to consider the detailed, almost endless, and emotional discussions that ensued after the *Luna 15* "landing." The debate revolved around the purported mission of *Luna 15*. There were three major guesstimates: a sophisticated soft lander with advanced instrumentation for surface analysis and photography, a sample return vehicle with some instruments remaining on the lunar surface for a modicum of longtime analysis, and a lunar rover. To the last I attached a likely weight of some 1,000 lb. The first choice faded out after some days for it did not survive the analysis of being classified as competitive to the *Apollo 11* landing. The other two choices were pondered over, argued about, endlessly analyzed, and ruminated about for months—no matter, the facts lay scattered about Mare Crisium where *Luna 15* landed and in the mental vaults of Soviet space officialdom but they were not revealed to the public.

Chapter 11: Return to the Moon—A Big Step

Completely missed by all but a few of us, were the launchings of *Cosmos 300* and *305*. *Cosmos 300* was launched on September 23, 1969, at 1407 GMT from Tyuratam into a 129.24 by 118 sm orbit, at 51.5°, and having a period of 88.24 minutes, while *Cosmos 305* lifted off from the same complex at 1414 GMT on October 22, 1969, into a 127.38 by 119.92 sm orbit, again at 51.5°, and with a period of 88.4 minutes. The first bird lasted in orbit for four days and the second for about two days.

Both of these satellites were lunar spacecraft attempts, which failed. They were certainly attempts to duplicate the *Luna 15* mission. Both of their translunar injection stages did not ignite—very unfortunate and very expensive failures. Two Zond boosters and two lunar payloads that together cost well over 200 million dollars brought no payoff at all. The next lunar probe would now be some time off to give sufficient time for thorough investigations and modifications to ensure success the next time around, whatever the mission to the Moon was intended to be.

The two failures disguised as Cosmos satellites ground the launching aspects of lunar missions to a complete halt for almost a year. That a gap in that program was to occur became rather evident after the *Cosmos 300* and *305* failures but the length of the gap was literally impossible to estimate. It turned out to be a long one. Not until September 12, 1970, at 1326 GMT (11 years to the day after the early *Luna 2* flight) did the Soviets announce that *Luna 16* had lifted off for the Moon. The announcement had the same words that were used in the statement on the *Luna 15* launch, and it was brief.

The flight to the Moon was almost identical to that of *Luna 15*, just 7 minutes longer, in fact, 101 hours and 55 minutes from translunar injection to arrival in lunar orbit at 2038 GMT on September 16. There was a midcourse correction early in the flight on September 13, it was of very short duration—6.4 seconds. The TASS pronouncements on the flight were very cautiously worded and each step was reported to the public only after successful completion. The first reported lunar orbit was circular at 68.4 sm above the lunar surface, at an inclination of 70°, and a period of 119 minutes.

The next day, September 19 saw still another maneuver to change the orbit to 65.9 by 9.3 sm, at 71°, with a period of 114 minutes—of course, a landing was in the making. Coolly and unflappable, TASS reported the landing, via radio and the Soviet press, as occurring at 0518 GMT on September 20, with selenographic coordinates latitude 0°41' south and longitude 56°18' east, in the Sea of Fertility [Mare Fecunditatis] (the eastern portion of the Moon, almost exactly on the equator). The landing program was complex and even unexpected in its format: first the retro-engine was turned on at a precise time for a specified burn. Then the spacecraft was permitted to "fall" freely in the lunar gravity field until the altitude of 600 meters was reached (about 2000 feet) whereupon the engine was turned on again and a "precision deceleration phase began…[with]… engine thrust mode…altered in conformity with the selected control program and the incoming information on the rate and altitude of descent."[86]

At an altitude of 20 meters (65 feet) the main engine was shut down and the subsequent operation was guided by low thrust engines down to an altitude of around 2 meters (6.5 feet). At this height above the surface all engines were shut down and the *Luna 16* spacecraft floated down to the surface for a very soft landing. Shortly after the landing the program of research was begun. That "research program" turned out to be drilling into the surface after appropriate photographs of the immediate vehicle area were taken and examined by Soviet scientists back on Earth in order to select the desired spot for such drilling. The actual drilling operation as described in several

---

86 No reference given in original document for this quoted text.

technical papers was no simple matter for temperature, drilling motor speed, and forces on the drill (among many other details) had to be carefully controlled.

However, the effort that had gone into the operation is better appreciated when the substance of several other papers is known. For instance, few in America know that full-scale replicas of *Luna 16* were constructed and thoroughly tested out in all aspects of the drilling operations (to say nothing of the usual system testing that ordinarily goes on). Various types of rock were drilled into on Earth with duplicate *Luna 16* apparatus and with diagnostic instrumentation used to check on every conceivable variable that could affect the mission. Of course, this was done in a vacuum chamber under simulated lunar conditions of temperature, lighting, and vacuum.

The success of the mission attests to the detail that had been given the simulation and diagnostics on Earth. The success was also due in large part to a point I've made again and again in this book; the unswerving perseverance of the leaders of the Soviet space program. Despite the failures of *Luna 15, Cosmos 300,* and *Cosmos 305* to perform this mission, the effort continued until a successful conclusion was reached for the initial sample return mission.

The sample, hermetically sealed, lifted off from the Moon, using the *Luna 16* landing stage as the launch platform, at 0743 GMT on September 21 after spending 26 hours and 25 minutes on the lunar surface. The return to Earth was much faster than the earlier trip to the Moon; it took just 69 hours and 43 minutes. Early reports from TASS described the possible landing area on Earth as very large, then as considerable. This caution in reporting was to cover contingencies that might have included an ocean landing as well as the remote, but not dismissible, possibility of a landing on another continent.

The actual landing was far better than initially expected; less than 300 miles northeast of Tyuratam from whence the spacecraft had lifted off to commence this really spectacular mission—how else could one describe it? The return flight was purely ballistic. There were no midcourse corrections nor did the vehicle carry engines for making any flight path alterations. The spacecraft had been aimed where the chosen landing area would be, and at burnout after lunar liftoff the velocity attained would have to be exactly that required for the Earth return. A few feet per second error and the spacecraft could miss the Earth entirely. A small difference in the aiming angle and bye-bye vehicle.

The procedure was essentially the mirror image of the flight to the Moon at the beginning of the mission. One "aims" the spacecraft where the Moon will be after a selected number of hours of flight so that the two—the spacecraft and the Moon—are on a near collision course. The spacecraft then arrives at the same place in space that the Moon is, adjusted so as not to collide but to orbit. If a direct landing is to be made then a real collision course is set, which is modified within the last few minutes by the retro-engines yielding a soft landing. Without retro-engines the landing is, as it was with *Luna 2*, completely destructive. All information from the latter type of landing must be obtained to be more exact, must be telemetered—before impact.

There was one significant difference in the trans-Earth trip from that of the earlier translunar flight. Going to the Moon was an adjustable event by virtue of the midcourse correction engine aboard the spacecraft. Coming back the liftoff had to be extremely exact for no adjustments could be made in the flight path. As reported above the flight was as exact as desired. To quote from *Pravda* of September 26, 1970:

All stages in the flight of the *Luna 16* probe, the flight to the Moon and in near-Moon orbit, the landing on the Moon, the collection of lunar soil, the lift-off from the Moon and the return to Earth, went off in conformity with the program and the computed data.

The search unit, provided with radar and aviation equipment, assured rapid detection and evacuation (that is, pick-up) of the reentry body of the Luna 16 probe with the lunar soil samples.

The vehicle entered the dense layers of Earth's atmosphere at 0510:00 (GMT).

The signal from the reentry vehicle transmitter was picked up at 0514:00 (GMT) and its descent by parachute was observed visually from the helicopters and aircraft of the search unit.

The vehicle landed at 0526:00 (GMT).

After the return to Moscow, the capsule with the lunar soil was extracted from the...container...maintaining the requisite sterilization.

The entire mission from liftoff to touchdown back on Earth had taken 280 hours and included 80 hours 40 minutes in lunar orbit encompassing some 41-plus lunar orbits.

The amount of lunar sample was surprisingly small, some 103 grams—a bit over 3½ ounces. The drill had gone about 14 inches into the lunar surface and had brought out a core about ½ inch in diameter. It was enough for a very complete analysis, nevertheless.

Three grams were eventually to find their way to the lunar scientists at NASA in the US.

The Soviets announced that the total weight of spacecraft that landed on the Moon was 1,880 kilograms—4,145 lb. Using that value and knowing the essential velocities for leaving lunar orbit for a landing, for going into lunar orbit initially, and for leaving Earth orbit for the translunar trip and also knowing the translunar trip time, I made some calculations to determine the weight in Earth orbit of the spacecraft before it left Earth orbit to go translunar. That is, the entire *Luna 16* spacecraft plus the entire propulsion apparatus to propel it to the Moon.

I determined a weight of 41,760 lb in Earth orbit and 10,200 lb for the *Luna 16* alone in translunar flight. These weights seem very reasonable in view of the values given several years earlier for the Zond booster in connection with the in-Earth orbit weight of the *Proton 4* satellite. My results required only a small improvement in the values given for *Proton 4*. Every booster ever built has been uprated over the years of its use. The Zond booster should be no exception.

The results of my calculations showed a translunar payload somewhat smaller than I stated for the *Luna 15* payload. The *Luna 15* calculation was made without any values such as those given by TASS for *Luna 16* so that the earlier estimate was really not too bad. However, the *Mars 2* and *3* spacecraft launched in May 1971 easily support the payload value in Earth orbit, that I gave for the Zond booster capability. They were each 10,000 lb and needed a substantially greater velocity to get to Mars than the Luna birds need to get to the Moon, therefore, the Earth orbit payload for the booster must have improved beyond the 42,000 lb used for the Luna spacecraft.

Photographs of the *Luna 16* spacecraft attest to its sophistication and the evident painstaking planning and design that preceded its use. Moreover, the craft was not small in size. No dimensions have been announced that I've seen but it appeared to be, perhaps, 20 feet tall and have a spread of about 24 feet at its foot pads. The success of the flight gave rise to numerous articles in the Soviet press, which included broad hints of more to come in both lunar and planetary flights. *Luna 16* was described as a testbed for a wide spectrum of new systems.

I have no doubt that these will show up over the next several years, since the discussions included return of soil samples from Mars, detailed exploration of the surface of Venus despite its furnace-like environment, and visits to other planets—not named, but I think Mercury, Jupiter, and Saturn are appropriate. In concert with Jupiter and Saturn explorations, the investigation of certain of their moons is rather a certainty. It is, in fact, easier to land on the moons than on the primary planet because of the lesser velocity requirements for those moons. But that is a long way off.

The Soviets were ecstatic over their lunar success and therein made a great case for unmanned, automatic vehicles for the exploration of space. The unmanned versus manned argument made its rounds again in the press, on the radio and TV, and throughout the science and engineering community. My opinion calls for continued exploration with automata until the capability exists for man to conduct the exploration. When the capability is at hand then we should do everything to promote the use of man in visiting the Moon, the other planets, and their moons. Let's face it, an unmanned spacecraft could land on some remote planetary body and end up sitting 100 feet from the discovery of this—or any other—millennia and miss it completely. The kind that a man sent into space wouldn't miss.

Success breeds success. Having established confidence in an activity we are all likely to make the most of the capabilities that led to the initial success. Ergo, it was not a long time to the next lunar exploration attempt. When the Soviets launched *Luna 17* on November 10, 1970, at 1443 GMT, I sent a note off to a friend, Bill Gregory, then managing editor of *Aviation Week & Space Technology*, and made the flat statement that *Luna 17* was carrying a lunar rover aboard. I had considered the case and had decided that it was just too soon for a *Luna 16* repeat performance. A total evaluation of spacecraft performance, drill operations, and sample analysis would take longer than two months. The cases discussed for the *Luna 15* mission had left the rover as the most likely mission after *Luna 16*. I had but a short time to wait to see if my judgment had been correct.

The translunar flight time to lunar injection was 104 hours and 49 minutes, over 3 hours greater than the *Luna 15* and 16 times for flight to the Moon. The Soviets were squeezing every drop of unnecessary velocity out of the lunar trip and putting it into payload. After 36 communications sessions—this bird's performance had even more attention than did *Luna 16*—and a midcourse correction on November 12, the *Luna 17* spacecraft was placed in lunar orbit on November 14 at 2150 GMT. The probe was in an entirely new lunar orbit; altitude a shade under 53 sm, inclination at 141°, and period of 116 minutes. This circular orbit was altered on November 16 so that with apogee still at 53 sm, a perigee was established at 11.8 sm. The period changed to 114 minutes. Again, a soft landing was imminent.

TASS lost little time in announcing the landing at 0347 GMT on November 17; 0647 Moscow time. At 1311 Moscow time word was on the air that the landing had occurred—safely! The broadcast stated that the landing was made with help "of a standardized landing stage." Evidently the same stage that had brought the *Luna 16* to the Moon's surface. The Mare Imbrium (Sea of Rains) was the site for this lunar visit and the landing was the farthest north any vehicle had yet visited; latitude 38°17' north and longitude 35° west in the extreme west central portion of the Mare Imbrium, just south of the Promontorium Heraclides. The latter forms the southern boundary of the Sinus Iridum (Bay of Rainbows).

At 0628 GMT, after lowering the double ramp for the vehicle, the rover *Lunokhod 1* (Moon car 1) was driven off of the landing platform by remote control from Earth. It was moved a bit over 60 feet from its landing stage. The TASS announcement went on to say that the *Lunokhod 1* carried a laser reflector system designed and fabricated in France. A lengthy discourse ensued on the difficulties of constructing such a vehicle and on the instrumentation that it carried for its lunar mission. It is interesting to note that although the wheel was chosen for locomotion over the surface, mechanisms were examined that "walked," slid, laid tracks (like a tank), and leaped. The final choice was an eight-wheeled vehicle, each of whose wheels consisted of three separate rims joined to a hub by spokes and connected to each other by cleats for better traction in the lunar soft surface. A fine mesh wire sheet was also laid over the circumference of the three rims to aid in traversing the Moon.

Lunokhod 1 was a real laboratory on wheels. It had instrumentation packed into all its available volume internally and then much was mounted on the exterior of the remotely operated spacecraft. There were six TV cameras aboard; the two forward cameras being high resolution systems and the four others, two on each side of the vehicle, low resolution cameras. The latter were used for taking panoramic photographs of the area in which *Lunokhod* landed. For the study of cosmic rays, detectors were mounted to determine the kind, intensity, and direction of travel for protons, electrons, and alpha particles. An X-ray telescope scanned the skies to seek out X-ray sources from other parts of this galaxy and from more distant places. An X-ray spectrometer irradiated the lunar surface and then examined the reflected nuclear particles to yield an analysis of the surface at that point. A penetrometer forced into the surface gave information on surface strength by recording the forces exerted on the vehicle chassis during the penetration. High gain and other antennas were used to telemeter all of the collected data back to Earth.

The top of the Moon rover opened in Sunlight to reveal an impressive array of solar panels on the inside of its cover, which supplied power to the craft. Presumably the top of the rover's body had some solar cells also. Mounted externally was a small solar panel used to supply electrical power before the main bank was opened at the beginning of a lunar day (14 Earth days). The side mounted cameras in concert with other instrumentation supplied data on the local vertical used in maintaining stable operation of the rover.

Since Lunokhod was in operation during the time *Venera 7* was on the way to Venus for a soft landing, a unique experiment was conducted. A solar flare occurring on November 19, 1970, was detected on *Lunokhod* and on *Venera 7* which was then almost 19 million miles from Earth. The data correlated very well.

All of this was certainly impressive but another revelation made after the landing was literally astonishing. A five-man crew that monitored and operated the rover were themselves monitored by a medical team. It had been established that in the excitement and enthusiasm of running the vehicle it was possible to make errors which could lead to "losing" the rover or endangering it in several ways. To lose an experiment costing $100 million or more because of a hasty or excited operator had to be prevented. The medical monitoring was a superb choice and showed unusual insight by the mission planners.

The rover crew was headed by a commander who had overall control of all operations; he supervised each move of the vehicle. A driver controlled all actual motions keeping in mind that each command to move had a 2.6 second delay built into it because of the round-trip time for a signal to get to the Moon and back to Earth. A navigator planned the path for the rover while the operator controlled antenna pointing so that the signal to Earth was always at a maximum. Finally, an engineer headed up a group of specialists for the conduct of all experiments, receipt of telemetry, and analysis of the new data.

When the rover was parked for the night at the end of the lunar day, it was positioned so that the passive laser reflector experiment could be performed. The reflector mounted on the top forward part of the rover was uncovered and the initial experiment conducted by the Krymsky Astrophysical Observatory on December 5 and 6. The distance to the Moon will eventually be measured, using this method, to about 3 feet. With some luck perhaps down to 1 foot.

In order that the temperature inside the rover be kept at reasonable levels a radioisotope thermal generator was mounted on the rear of the *Lunokhod*. It contained Polonium-210, which emits alpha particles during its radioactive life. Alpha particles do not penetrate very deeply so that the containing structure prevented the radioactivity from any sort of interference with the experiments. It is evident that the amount placed aboard was quite conservatively estimated for it lasted far longer than the mission was originally intended to last.

Lunokhod was intended to move at two different prescribed rates, the second of which was twice the first. Neither velocity was ever announced. The vehicle could be programmed to move a given distance and then come to a stop. This was used on several occasions. Although the rover was designed to overcome tilting to one side by 14°, it actually overcame tilts of 28°. It was designed to move up slopes of 10-15° but managed 27° slopes.

By the end of three lunar days on the Moon, a strip of some 12,000 feet long and 500 feet wide had been surveyed and 14 chemical analyses had been conducted. Moreover, a great deal of physical analysis of the surface had also been accomplished. I had supposed that *Lunokhod* was a vehicle in the 800 to 1,000 lb class as I studied its deeds. I was awakened to the facts when the Soviets announced that *Lunokhod* was much bigger, 1,667 lb. It was 62.4 inches from center rim to center rim across the vehicle's front, 85.6 inches from the front of the forward wheel to the rear of the last wheel, it stood 62.5 inches high to the top of its cover from the "ground," and was 83.9 inches across the cover from a frontal viewpoint. That made the rover higher from the ground than my Mustang, wider than the Ford product, and more than half as long. Its weight is more than half that of the car too. Being "a bit advanced" over Earth transportation design the rover had disc brakes, electromagnetically operated, and individual drives for each of the eight wheels. That all adds up to a lot of rover on the Moon. Far more than anyone expected and an excellent example of the optimum use of weight and instrumentation for this superb mission.

The mission for *Lunokhod 1* was set as three months but not loudly proclaimed as such until that time had passed. Many believed that the mission planners would have been satisfied with less. No matter, the *Luna 17/Lunokhod 1* mission was a fabulous success for the rover was active for 10½ mouths, easily surpassing all expectations. The rover had traveled 5.5 miles, covered over 800,000 square feet of the lurain, transmitted 20,000 first rate photographs and 200 panoramas of the lunar surface. The chemical composition of the surface had been determined at 25 separate locations and physical and mechanical properties of the lurain had been determined at 500 points. The rover was parked in a position to permit the laser reflector experiment to continue indefinitely. It should be clear that the only limiting factor here was the depletion of the radioisotope, Polonium-210, otherwise the mission could have continued without end. It is entirely possible that sometime in the future the power source could be replaced and the rover reactivated. Who knows?

I find one further set of remarks that are pertinent to the explorations by *Lunokhod 1*. Quite possibly the extensive and inordinately detailed survey performed by the rover may just have been an exercise through which to determine the capabilities of the vehicle; however, that type of surveying is required when a lunar base is being planned. Since that kind of information is not likely to be broadcast in advance of the event, I can only surmise at this point that the *Lunokhod* survey was made with such a purpose in mind for the future. And only the future will tell.

These large lunar spacecraft were not being launched at the rate the Soviets had used for the earlier birds—*Luna 4* to *Luna 14*—for the next launching in this series did not appear on the scene for almost one year. It wasn't until September 2, 1971, at 1340 GMT that *Luna 18* was launched and it headed for approximately the same area of the Moon that *Luna 16* had helped to explore. There were two midcourse corrections on September 4 and 6 and arrival at the Moon took place at 2326 GMT on September 7.

What really surprised me here was the length of the trip from translunar injection to lunar arrival in orbit. The trip time was 128 hours and 29 minutes, exceeding by a whole day prior trip times. A trip that slow is essentially on the verge of not getting there at all, or, conversely, of just barely having enough velocity to escape from the Earth for the intended lunar expedition. The spacecraft performed 54 turns around the Moon before the soft-landing attempt was made.

Not long after 0700 GMT on September 11 the landing effort began. However, the fates were unkind for communications ceased at 0748 GMT with no further reported response from the spacecraft. It was announced that the landing attempt was being made in the lunar highlands and I suspect that *Luna 18* landed on a steep slope, tumbled, and was lost. Because of the remarks, released by TASS, that bespoke a last moment "unfortunate occurrence" I think my estimate of the event is likely to be correct. *Luna 18*'s location on the surface is at latitude 3°34' north and longitude 56°30' east. There was no further discussion by the Soviets of *Luna 18* but the remark about a highland landing suffices to say that the mission was a repeat of the *Luna 16* mission without the latter's success.

The validity of my impatient remarks about long times between launchings got all shot up in just a few days after the *Luna 18* failure to land safely on the Moon. *Luna 19* was launched on September 28, 1971, at 1000 GMT, a Tuesday, and just 17 days after the *Luna 18* mishap. Again, there were two midcourse corrections on September 29 and October 1, respectively, and arrival occurred at 2122 GMT on October 2, some 106 hours 15 minutes following translunar injection. The first announced lunar orbit was circular at 87 sm, at an angle of 40°35', and with a period of 121 minutes 45 seconds. There was a minor orbit change on October 6 and then about December 2 a substantial change was announced yielding an orbit of 239 by 47.8 sm, at an angle of 40°41', with a period of 131 minutes.

Of course, during the interim after lunar arrival my colleagues and I speculated all over the place about the *Luna 19* mission. Evidently it is a long interval photographic mission, for as late as March 1972 it was still operative and returning photos of the surface. The specific reason for further detailed photographs appeared to become evident in late March 1972 when Dr. Alexandr P. Vinogradov, a Vice-President of the Academy of Sciences of the Soviet Union, announced that an attempt would be made to retrieve samples of the lunar surface from the farside of the Moon.

Luna 19 could have collected some high-resolution pictures to aid in choosing the site for such a mission. Insofar as I can tell, it was still operative in lunar orbit in mid-May. If so, it established a record as the longest operating lunar satellite of any complexity. For a farside landing the Soviets would need a communications relay. Total programing, in advance, of the lander; landing, sample retrieval, liftoff, and return to Earth is too big a bite and too demanding a task for such a mission. There has been no additional word about *Luna 19* since March 19, 1972.

One could hardly expect the Soviets to back off from this class of missions after having made such a large commitment to it. The announcement of the flight of *Luna 20* on February 14, 1972, at 0320 GMT, according to TASS was as planned and *Luna 20* arrived at the Moon on February 18, having experienced one midcourse trajectory correction on February 15.

The usual cautious procedures in spacecraft checkout and careful orbit determinations took place for the next three days. I wondered what part *Luna 19* played in all this, but to no avail for there has been not the slightest hint of its participation, if any. The landing took place on February 21 as reported in *Pravda* as follows:

> At 2219:00 Moscow time (1919 GMT) on February 21, 1972, the *Luna 20* settled on the Moon's surface at a point with the selenographic coordinates 3°32' N and 56°33' E. As has been reported, the Luna 20 was injected into a selenocentric circular orbit on 18 February. The motion of the probe was corrected on 19 February, with the result that it was transferred into an elliptical orbit with

maximum altitude above the Moon's surface of 100 km and minimum altitude of 21 km.

To assure the probe's landing at the designated area on the Moon, the main retro engine was switched on at 2213:00 (Moscow time) on 21 February. The power plant was switched off after 267 seconds, and the probe fell freely to an altitude of 760 meters.

The probe then descended in a controlled descent mode, during which the thrust of the main engine was changed by using the control system. Starting with the altitude of 20 meters above the surface of the Moon, deceleration was accomplished by using the low-thrust engines.

The landing point of the Luna 20 is on a portion of the lunar continent adjacent to the northeastern neighborhood of the Sea of Fertility.

Luna 20 had landed less than 1.13 miles from the site of the *Luna 18* spacecraft that had made a bad landing. And having landed in the lunar highlands its mission became evident. No one would land a lunar rover in the highlands. This was a repeat of *Luna 16* and success looked quite certain, now.

The stay on the surface was 27 hours and 39 minutes; at 2258 GMT on February 22, *Luna 20*'s return vehicle was launched and headed for home. Arrival at Earth occurred at 1912 GMT on February 25. The whole round-trip time from launch to landing back on Earth took 279 hours 45 minutes (11 days, 15 hours 45 minutes). The return to Earth took place in a fashion somewhat more spectacular than the Soviets would have preferred. Upon entry into the atmosphere the return capsule descended through a howling blizzard. To aid the recovery helicopters in tracking the return vehicle a large quantity of thin metallic needles was released after the reentry procedure. These needles provided an excellent reflecting mechanism for the aerial search radar—and presumably for search radar on the ground directing the recovery craft. The sample return spacecraft came down less than 300 miles northeast of the Tyuratam launch base from which it was originally launched. If, indeed, Tyuratam was used as the "aiming point" for the return, then the miss represented an error of less than 0.07°. For a 250,000 mile shot such accuracy is utterly remarkable.

As expected, the *Luna 20* sample was quite different from that obtained with the *Luna 16* spacecraft but discussions of the scientific analysis associated with the samples is best left to technical publications. Suffice to say that the Soviet Union is pleased with the results. We can expect to see more such flights, and with retrievals from the lunar farside they will become spectacular, indeed; no matter that the fickle public has lost interest. The advance of science needs only advocacy and funds. The Soviet Union is supplying both for their scientific/engineering community.

Samples of *Luna 16* and *Luna 20* retrievals have been given to the scientists of NASA's lunar laboratory at Houston, Texas. In return the Soviets have been given some Apollo sample returns.

The *Lunokhod 1* mission in 1970 was considerably more successful than had been anticipated by Soviet engineers and scientists. It operated for so long beyond its design life that an extended program of investigation was devised and used. This implied that a similar vehicle could, with appropriate alterations in the instrumentation package and carrier design, accommodate additional experimentation on the lunar surface. After an interval of time during which modifications were made *Lunokhod 2* evolved. *Luna 21* carrying *Lunokhod 2* was launched at 0653 GMT on January 8, 1973; the first launching of that year.

From a low Earth orbit the spacecraft proceeded to the Moon, arriving on January 12 and went into a 68.4 by 55.9 sm orbit, inclined 60°, with a period of 118 minutes. After 41 selenocentric orbits and a final change of perilune to about 10 miles, *Luna 21* landed at 2235 GMT on January 15 in Le Monnier crater about 3½ miles from Montes Taurus; selenographically at latitude 25°51' north, longitude 30°27' east. The landing site was some 140 sm due north of the landing site of *Apollo 17*. Le Monnier crater is 30 sm in diameter and is located at the eastern rim of the Sea of Serenity. *Lunokhod 2* came off of its landing platform for a short test run on the surface at 0114 GMT that (Earth) morning.

The 1,852 lb (840 kg) eight-wheeled spacecraft, almost 200 lb heavier than its predecessor, carried instrumentation with which to conduct both chemical and physical analyses of the lunar soil. It carried a magnetometer, a complement of video cameras, a drill, a French-built corner reflector for laser studies, astrophotometers for various light spectra measurements, a penetrometer, and several types of nuclear particle measuring devices. Video camera equipment was improved so that single pictures could be transmitted in 3 seconds (*Lunokhod 1* video pictures took 30 seconds each for transmittal). The astrophotometer was used to measure ultraviolet spectra, the zodiacal light, and lighting in several stellar regions. The astrophotometer and the magnetometer had not been aboard *Lunokhod 1*. Nor was the Rubin 1 receiver, a photon detector used in conjunction with Earth-based lasers to determine the location of the vehicle and its direction of motion on the lunar surface. Conservative, as always, the controllers permitted *Lunokhod 2* to traverse only 3,800 feet (1,160 m) during its first 100 hours on the Moon.

Discovery of a smooth slab of rock with no pockmarks was made about a mile from the landing site. The unusual rectangular geometry of the 1 meter long rock and the fact that application of 1,470 psi applied using *Lunokhod* left only the barest impression in the dust coating on the rock gave rise to additional and prolonged study of this specimen. Evidently the lack of marks indicated an age for the rock much less than its surroundings.

Earlier while mid-lunar day lighting obscured the surface preventing Earth-located controllers from making any but small movements, still a number of experiments were conducted on January 18. Solar X-ray radiation was measured as was "lunar sky" illumination, and mechanical properties of the surface soil were examined. Several video panoramas were received at the Earth control station; these also permitted an appraisal of the condition of the landing stage of *Luna 21*. The following (Earth) day, *Lunokhod 2* headed south toward the crater rim, stopping to examine surface conditions, determining ease of traverse over the surface and frequently taking video pictures of objects both nearby and afar.

During the next two lunar days, *Lunokhod 2* traveled over 17 miles exploring, photographing, and conducting both chemical and physical analyses of the surface. Its magnetometer was activated on numerous occasions and further studies were made of ultraviolet spectra, the zodiacal light, and solar radiation. Some of the soil analysis findings showed a wide variation in the iron and aluminum content of the soil as a function of location. A survey article covering the findings of the *Lunokhod 2* mission can be found in the May 1974 issue of the British Interplanetary Society magazine, *Spaceflight*.

The lunar rover crossing some very rough and hazardous areas ran into the statistics of accidents. After the start of the fifth lunar day, on May 9, somewhere in the rough mountainous sections south of its landing site, *Lunokhod 2* had an accident not described in the press by the Soviet Union. At the end of the fourth lunar day the rover had traveled 22.44 miles, while at the conclusion of its mission it had moved a total of 22.94 miles. Hence the accident occurred shortly after day five had started. Only one-half mile of motion had ensued during the fifth day. It is known that the vehicle had changed its elevation on the surface by more than 1,000 feet, which infers a climb into the foothills of Montes Taurus or into the rocky slopes of the rim of Le Monnier crater or, possibly, on the slopes of a lesser crater within Le Monnier.

The Soviet Union's communique had stated that the mission had been completed after four months' operation of the vehicle but the announcement on Sunday, June 3, 1973, was made almost a month after mission actions had ceased. Part of the interval may well have been used to attempt to recover from the mishap, without success. When the Soviet Union was trying to soft-land a spacecraft on the surface of the Moon in the mid-1960s, with a particular eye to laying claim to a first, a launch occurred every few months until that goal was attained in early February 1966.

By the mid-1970s the pace had slowed considerably for many reasons; there was no longer a "race" to the Moon, spacecraft had grown in size, sophistication, and cost giving rise to longer intervals for conception, design, fabrication, and testing. The testing, in fact, had become an intricate procedure rivaling the complexity of the projected experimentation itself. This was not unexpected. Nevertheless, the flight of *Luna 22* was a long interval away, almost a year after that of *Luna 21*. It occurred on May 29, 1974, at 0856 GMT.

Leaving the now familiar low Earth orbit, the spacecraft arrived at the Moon on June 2, but no descent to the surface was forthcoming; this intent had been announced in the original communique on the launching. The initial orbit was circular at an altitude of 137 sm, inclined 19° 35', with a period of 130 minutes. The major mission appeared to be photography of the lunar surface and this was supplemented by the usual array of studies of cosmic rays, plasma concentrations near the Moon, meteoritic material near the Moon, and gamma ray spectra analysis of the surface taken in conjunction with the photography sessions. High-resolution photography was obtained by first lowering the spacecraft's perilune to 15.5 sm. A radar altimeter aboard was used to determine surface profiles, also in conjunction with the photography; this permitted formation of a surface relief map of the Moon for selected areas. It could also contain implications for future landings, manned or otherwise.

Over the next several months the perilune and apolune for *Luna 22* were altered many times in concert with the conduct of various experiments. Small changes in orbit inclination were made as well. Table XII contains TASS released data on the orbit changes. *Luna 22* was fully operational five months later when *Luna 23* arrived in lunar orbit on November 2. More than 1000 communications sessions had been held with *Luna 22* in those five months.

On August 24, 1975, the spacecraft, under the plan of additional programming, lowered its perilune to 18 miles and obtained a new series of video pictures of the surface. After the picture-taking perilune was raised once more to a new altitude. In the last bulletin on *Luna 22* issued on September 4, 1975, TASS noted that the spacecraft after more than 17 months residence in lunar orbit was still operating.

Table XII: *Luna 22* Lunar Orbit Parameters

Date	Apolune	Perilune	Inclination	Period
Jun. 2, 1974	137	137	19°35'	130
Jun. 9, 1974	152	15.5	—	—
Jun 13, 1974	185.8	112.5	—	—
Nov 11, 1974	893	106	19°33'	192
Apr 2, 1975	876	124	21°	192
Aug 24, 1975	—	18.6	21°	—
Aug 24, 1975	799	62	21°	180

Apolune, perilune in miles; period in minutes.

A pattern of sorts had evolved with the Soviet's large Luna spacecraft and fittingly enough could be described as a troika. In two cycles the repetition of the troika has been duplicated exactly provided the intervening failures are omitted. The troika is, in order of their successful occurrence, a sample mission, followed by a rover, which in turn is followed by an orbiter. Thus;

Luna 16   S   *Luna 21*   R

Luna 17   R   *Luna 22*   O

Luna 19   O   *Luna 24*   S

Luna 20   S

S = sample return mission; R = lunar rover; O = lunar orbiter.

Luna 16 was a repeat of the failed *Luna 15,* which attempted to offer competition to the *Apollo 11* manned lander. Later on, the successful *Luna 24* was a repeat of the *Luna 23* mission, which had landed but damaged its drilling equipment in the process thus cancelling the mission. *Luna 18* was a sample return mission that crashed right at its landing, evidently coming down on boulders or a very steep slope.

If this serialization is not just a coincidence, then *Luna 25*, whether a late 1976 or early 1977 mission, should be the carrier for *Lunokhod 3*. How long such troikas will continue to appear is difficult to surmise. The future should bring a lunar base, unmanned at first but eventually manned. The time scale for the base will be a strong function of the availability of a Saturn V class booster and/or a reusable space shuttle; either of these are likely to be coupled with a translunar-trans-Earth tug.

Luna 23, the intended sample return mission was launched at 1430 GMT on October 28, 1974, the third time that two large Luna spacecraft had been launched in a single year. (It should be pointed out, however, that in addition to *Luna 15* in 1969, *Cosmos 300* and *305* were launched in September and October 1969 but both failed to leave Earth orbit, neither having received Soviet identification as intended Luna spacecraft. Both decayed and disintegrated in the Earth's atmosphere.)

Upon its arrival at the Moon on November 1, *Luna 23* went into a 64.5 by 58.3 sm orbit, at an inclination of 138°, and with a period of 117 minutes. *Luna 23* transferred its orbit to one of 65.2 by 10.6 sm, and then on November 6 at 0537 GMT after a 6 minute landing procedure came down in the southern part of Mare Crisium at latitude 13.5° north, longitude 56.5° east. The drilling apparatus, designed to retrieve an 8 foot long continuous sample was damaged on landing.

Because the mission was designed for that specific purpose alone a very abbreviated program encompassing engineering studies of the operation of the landing craft was made and the mission totally concluded on November 9, 1974. Of the ten *Luna* spacecraft, *15* to *23* inclusive, eight were landing mission and of these five were totally successful while three failed. *Luna 15* apparently had a control failure and hard-landed on the surface. *Luna 18* landed in a very hazardous area and was lost as a result, and *Luna 23* having damaged its major equipment during landing could not complete its mission. Conquest of the Moon by man or by sophisticated robot has hardly begun.

12

THE PUZZLES

In any study of so complex a system as the Soviet space program, only the most fortuitous circumstances would present a completely clear picture to even the most diligent of researchers. That improbable series of events did not occur in this case and so there remains a number of launchings whose objectives remain a puzzle to this day. The information on them is very limited. These several spacecraft were launched and announced in the usual manner but with virtually no information after the initial launching events.

In stating that these spacecraft offer a puzzle, I intend that they are distinct and separate from what has been recognized as various failures for other spacecraft that did reach Earth orbit but which did not complete the remainder of their missions; e.g., the disguised *Cosmos 111*, an intended Moon probe; *Cosmos 96*, and *Cosmos 167*, intended as Venus spacecraft; and *Cosmos 41*, an early attempt at establishing the *Molnyia 1* communications satellite system.

In a related category, one should include the 74° inclination satellites whose several purposes have never been adequately explained: those put up in clusters of eight appear to be navigation or communications related in the military sphere while those placed in orbit singularly may be ELINT or, again, navigation satellites.

The specific mysteries I am referring to are listed below:

Spacecraft Launch Base	Launch Date & Time (GMT)
Cosmos 379 Tyuratam	November 24, 1970 / 0514
Cosmos 382 Tyuratam	December 2, 1970 / 1634
Cosmos 398 Tyuratam	February 26, 1971 / 0506
Cosmos 434 Tyuratam	August 12, 1971 / 0429

Still obscure even at this date—1975—are the missions for these four spacecraft. There is, in fact, some reason to believe that possibly two different missions are represented in these four satellites, for *Cosmos 382* has a somewhat different history than the other three.

Human nature is such that we feel most comfortable when some current event can be seen to fit in with what we believe to be a series of past and connected happenings. And based on those, presumably historically related episodes we make analyses and not infrequently predictions. Frequently the evidence is pretty thin, but we go out on a limb because answers to various questions are almost demanded in one area of society or another. Such answers are hedged because of the uncertainty associated with the blanks in data. On the other hand, there are those who will attempt to answer many questions hastily while assuming the position of knowledgeability where the latter does not really exist. And, finally, in contrary fashion, some answers come forth, incorrectly, because an occurrence (in our case spacecraft launching) is so well disguised.

Early interpretations become misinterpretations. So, it was with *Cosmos 379*, the first really puzzling satellite from the Soviet Union in a long time. This spacecraft was launched from Tyuratam on November 24, 1970, at 0514 GMT into what looked like a very standard reconnaissance satellite orbit at 157 by 123 sm, at an inclination of 51.6°, and with a period of 88.7 minutes. Nothing appeared to be unusual initially. On December 7, the very knowledgeable writers in *Aviation Week & Space Technology*, certainly the bible of the aerospace world, commented on this launching in their "Industry Observer" column:

> Major Soviet space failure appears to have occurred November 24 with launch of Cosmos 379 from Tyuratam. Proton booster was used but the payload's 158 x 124-mi. altitude is far lower than that of previous 17-ton Proton payloads. Booster casing decayed after only two days in orbit and no signals from the payload are being received. Observers speculated that Cosmos 379 could have been failure of Russia's Statsionar synchronous orbit communications satellite.

Several things should be pointed out here. First, the "Proton" booster is the Zond three-stage booster not the earlier two-stage launch vehicle. (The first three Proton satellites were launched by the two-stage booster while *Proton 4* was launched by the three-stage Zond booster, the outgrowth of the Proton booster. Only *Proton 4* was 17 metric tons, the others were 12 metric tons.) This difference has been inadvertently missed by *Aviation Week & Space Technology*; it has been some years since the third stage was put into use.

Second, the disclaimer in the paragraph "appears to have" would ordinarily have allowed elbow room for error but as early as the 17th orbit, about a day after the launching, *Cosmos 379* made a radical change in its orbit. It moved to a 748 by 121 sm, 98.73' orbit. By orbit 21 it had made an even more radical change to an 8718 by 108.3 sm, 260' orbit. There were, in the course of events, reports that taped voice signals were being beamed from the spacecraft.

There were no launch windows open for lunar or planetary flights at that launch time. The orbit parameters seemed inappropriate for any recognizable space station experiments. Almost all new space-related experiments are conducted from Tyuratam and certainly all of the important ones come from that launch complex. The orbit inclination could be a link with manned spaceflight. On the other hand, there were many reconnaissance satellites at or very near that inclination. The maneuvering could be related to manned or to reconnaissance flights.

One might make something of a case for relating this flight to a large astronomical observatory, where the high apogee allows for a slow rate at that altitude and therefore eases the difficulty in pointing and the low perigee allows for easier retrieval of data. I don't believe it, however. There is no fun in not paying some attention to the playing of games, guessing games, so I will, despite the lack of hard evidence of any sort. The experiment could be related to manned spaceflight and possibly to a Soviet space shuttle. The rocket engine used to change orbits is probably some five to ten times as large as the Soyuz maneuvering engine. Say five times the thrust (4,410 lb) just to play it in a conservative engineering manner.

If not this particular choice then I'll join the ranks of the baffled, at least for the time being.

As if this launching weren't a hard enough kernel to gnaw on for a while the Soviets made it more interesting in very short order. On December 2, just eight days after the *Cosmos 379* launching, *Cosmos 382* was launched at about 1630 GMT from Tyuratam into what was announced as a 3132 by 199 sm orbit, at 51°35', and with a period of 143 minutes. Though *Cosmos 379* and *382* had similar orbit inclinations, they were not in the same orbit plane and hence not positioned for a rendezvous and docking exercise.

I demurred on my exact calculation for the liftoff time because there was an orbital maneuver completed before the initial indications given in the first *NASA Prediction Bulletin*. The existence of the maneuver was clear enough for the zeroth node (first crossing south to north) of the equator was given as longitude 335° instead of the usual 346-348° for the 51° inclination satellites. I am fairly certain, that knowing these facts, a computer program could be devised to determine the true launch time.

In any event the satellite went through a series of further maneuvers after attaining the announced orbital position and they looked like this:

By Orbit No.	Apogee	Perigee	Period	Inclination
	(Miles)		(Minutes)	(Degrees)
30	3135	250	143.5	51.48
45	3148	1002.8	158.93	51.55
48	3155	1599.6	171.03	55.87

Finally, somewhere downstream (i.e., much later) and still in that position by orbit number 2584 on October 5, 1971, a last adjustment to 3232 by 1519 sm was made while maintaining the same period and inclination as at orbit 48. The notable item about these several maneuvers was the inclination change of 4.3° at about orbit 48. In terms of velocity and propellant expenditure that inclination change was expensive; it could cost as much as 1,800 fps. That is a good deal of propellant for a large payload—the general interpretation for *Cosmos 382*.

A major question that arises concerns the possible connection with the *Cosmos 379* spacecraft. Similarities exist in maneuvering many times, in the original orbit inclination, in using the same launch vehicle—the Zond booster—and in baffling all the non-Soviet observers. The mystery still exists. One might conclude that the maneuvers were simply a byproduct of the test of a new propulsion system together with appropriate guidance and control wherein the altitude changes per se reveal nothing. That too remains to be discovered.

No good having a mystery that doesn't thicken once in a while. The Soviets obliged on February 26, 1971, at 0506 GMT by launching *Cosmos 398* into a 172 by 122 sm orbit, at 51.63°, and with a period of 88.9 minutes. This was very much like the *Cosmos 379* orbit. Sure enough, by orbit 39 on March 1, the satellite maneuvered to a 6748.5 by 124.2 sm orbit essentially at the same inclination but with a new period of 215.93 minutes. Again, the Zond booster was used and again the Soviets made no announcement after the initial TASS bulletin on the launching. Further data on this launching is lacking except for a particular speculation.

Some items have appeared in the technical press stating that *Cosmos 382* had made maneuvers whose velocity increments coincided with those necessary for injection into a lunar orbit (i.e., at arrival) and then for a trans-Earth injection from the Moon. These maneuvers were made in Earth orbit. This commentary is in distinct contradiction to the calculations that I have made. The movement to higher orbit for *Cosmos 382* was made in a separate series of maneuvers none of which included a Delta V that satisfied either the lunar insertion Delta V or the trans-Earth Delta V.

There were more than four maneuvers plus the fact that the total velocity incrementally added was close to 6000 feet per second. Hence no single maneuver included a velocity increment to cover the lunar insertion or trans-Earth events. One must conclude that the original information source was faulty. I emphasize again that my calculations were based on data published in the *NASA Prediction Bulletin* (i.e., SPADATS data) and consequently are well supported and therefore I believe to be quite correct.

One might assume that the aforesaid information source was well aware of the fact that the 6,000 feet per second velocity increment in total was indeed enough to cover the lunar maneuvers discussed but was not aware of the separate maneuvers made by the satellite and that he either guessed or made up a story to fit the data. In either case he was wrong. The implications of real practice for such maneuvers are very great and clearly says a great deal about the existence of a Soviet lunar program.

In the years following these events, nothing has happened to further support evidence of a Soviet manned lunar program. While I am inclined to believe that the Soviet Union yet has an eye on the Moon, their intentions therein involving man are yet to be revealed. After a lapse of six months another member was added to the triumvirate of spacecraft when *Cosmos 434* was launched from Tyuratam at 0429 GMT on August 12, 1971. The orbital ballet was initiated once more as the spacecraft, in four separate steps, moved to an orbit almost 7400 miles high at apogee, with a perigee finally at 112 miles.

The timing of the launching seemed to place it in a class with *Cosmos 379* and *398*, since they too were launched in the early hours (both about 0500 GMT) whereas *Cosmos 382* was launched much later in the day at 1634 GMT.

The well of knowledge has run dry insofar as good, solid data is concerned on these satellites. Speculation, however, abounds; the literature is full of thinly based calculations claiming these spacecraft to be tests of a lunar module or of boilerplate Salyut type space stations (one might consider why a boilerplate model after an operational station has been launched and used). I am strongly inclined to agree that the spacecraft are involved with future manned flights and hence a new program—this particular speculation remains to be confirmed. Time will unfold the facts.

13

The Future
Soviet Space Activity for the Next 10 Years

Interplanetary (Unmanned) Activity

The Soviet Union's space program includes plans for the investigation of all of the planets of the solar system through the use of unmanned interplanetary spacecraft. They have announced their intentions to launch spacecraft on the grand tour type of mission, stating that their plans call for investigation of Jupiter, Saturn, Uranus, and Neptune. They explicitly stated an intent to launch spacecraft for that mission, "in early October of 1978," (*Soviet Life*, circa 1968). This does not preclude earlier launchings as well, or ones later in 1979. Because these kinds of experiments are inherently very time-consuming, and since the opportunities to attempt them are presented on rare occasions, I expect that the Soviet Union will make an extensive effort for these launchings. I expect:

- Several spacecraft to be launched in the grand tour mission in each of two and possibly three different years; 1977, 1978, 1979.
- Each year for the grand tour ought to see a minimum of two spacecraft launched, quite possibly three spacecraft; and conceivably there may occur of the pertinent years as many as four launchings in each.
- The main spacecraft will carry auxiliary spacecraft to:
    1. Enter the Jovian or Saturnian atmosphere, or
    2. Orbit Jupiter or Saturn, or
    3. Orbit one of the natural satellites of Jupiter or Saturn, or
    4. Attempt a soft landing on one of the atmosphere bearing moons of Jupiter or Saturn, or
    5. Float a payload in the upper atmosphere of Jupiter or Saturn.
- The booster to be used will be the Soviet Union's launch vehicle having a first-stage thrust of 10 million pounds or more. Although the first flight attempt using this booster was reported to have ended in a disastrous explosion on its launch pad (*Aviation Week & Space Technology*, November 17, 1969) we can reasonably expect it to be thoroughly operational before the grand tour opportunities present themselves.
- Since the spacecraft trajectories will be receding from the Sun and the distances for sending data commencing at hundreds

of millions of miles from Earth, transmitters will have output power greater than 100 watts from auxiliary spacecraft and may run to a kilowatt or more from the main spacecraft.
- The data situation may lead to data relay: auxiliary spacecraft to main spacecraft to Earth.
- Efforts will be made to obtain data directly as well from the auxiliary spacecraft; spacecraft solar system geometry will be an important consideration in such cases.
- Power sources will be either radioisotope thermoelectric generators (RTG) or nuclear reactors. The latter will likely be an improved version of the Romashka (daisy) reactor.[87]
- Experiments will be aboard to:
    1. Analyze planetary atmospheres for constituents.
    2. Determine temperature profiles elemental or compound.
    3. Determine pressure and density profiles.
    4. Analyze the Jovian radiation belt.
    5. Examine atmospheres of moons like Ganymede, Callisto, or Titan.
    6. Examine the rings of Saturn by attempting some sort of spectral analysis.
    7. Determine the nature of the great "red spot" in the Jovian atmosphere.
    8. Search for other as yet unknown moons orbiting either Jupiter or Saturn, or Uranus or Neptune.
    9. Photograph everything in range of the cameras; primaries, satellites, rings of Saturn, the Jovian red spot,

---

[87] The Soviets have stated that they have developed a space-flyable nuclear reactor they call Romashka.

the Sun and, perhaps, some of the asteroids if they can be detected well enough for photography.

10. Less likely but not a dismissible possibility is photography during descent into the Jovian and or the Saturnian atmosphere with video transmission to the mother spacecraft for relay to Earth. In this same category is a landing on Titan.

An auxiliary and interesting aspect of the grand tour mission is the need to use the Soviet large booster (Saturn V class) for this mission. The Zond booster—used to launch *Zond 4, 5, 6, 7*, and *8*—can launch about 1,500 lb on to an initially Jupiter-bound trajectory. Moreover, staging must occur during injection to get this much payload because of the large velocity required. Now knowing the Soviet Union's space program history, this (small) size of interplanetary payload has not been launched since *Venera 1* (1,428 lb) in February 1961; 1,500 lb is far too small for current requirements and desires.

I made a brief examination of the propulsion injection requirements for a 1977 grand tour and assumed only a minimum injection velocity increment of 23,000 fps (48,000 fps total ΔV). I used a specific impulse of 350 lb-sec/lb and divided the velocity increment between two injection stages.[88] The results are as follows:

Trans-Jupiter Injection Stage, Approximate Weights for the Grand Tour Mission

Stage 1:   weight   ≈250,000 lb (in Earth orbit, see Figure 6)
           ΔV_1 7,000 fps
           Propellant   115,070 lb

---

[88] G. V. Petrovich, "Home of Cosmonautics," *Vestnik Akademii Nauk Kazakhokoy*, no. 10, 1967. Covers the description of an upper stage engine with a specific impulse of 353 sec, etc.).

	Structure, etc.	10,000 lb
	Payload (Stage 2)	124,000 lb
Stage 2:	weight	124,000 lb
	ΔV_1	16,000 fps
	Propellant	94,000 lb
	Structure, etc.	10,000 lb
	Grand Tour payload	≈20,000 lb (see Figure 6)

It is important to note that the velocity increment cannot be attained by burning all the propellant in a single stage. A major factor in this is the Soviet Union's continued lack of use of high energy propellants. If 10 million pounds is indeed the first-stage thrust of their large booster, then the numbers given above are good approximate values for the injection procedure. The payload of 20,000 lb is in keeping with a desire to obtain useful scientific results, to ensure sufficient redundancy so as to enhance reliability in view of the long trip time, and would allow most of the experiments listed above to be carried out. The approximation is conservative in that a more accurate analysis would yield lower structural weights. Therefore, any midcourse correction weights would be covered in a refined analysis without adding to the weights given above. To understand Soviet booster payload capabilities, see Figure 6.

Up to 5,000 lb of the 20,000 lb payload could be used to satisfy the items of Table VIII. Although a larger package would be better for items 2 or 3 of Table VIII, a 2,000 lb payload could be orbited around Jupiter using part of the 5,000 lb to supply an 8,000 fps velocity decrement to retro into Jovian orbit. This also assumes a specific impulse of 350 sec in the retro propulsion system. If the specific impulse is as low as 300 sec then only 1,700 lb can be orbited in this case. This small payload would probably have an RTG-based power system with a power output of 100-250 watts for a long period. A parabolic antenna would be a necessity for this payload. The antenna would be between 2 and 5 meters in diameter to yield an adequate gain.

It would not surprise me to see an antenna larger than 10 meters on the main grand tour payload. A fascinating possibility for the destination of one of the grand tour payloads (GTP) is an orbit around Uranus with subsequent spectral analysis and photography of the planet and any of its five moons. A 5,000 lb payload could be orbited using the 15,000 lb GTP that remains after dropping off the small payload to orbit Jupiter. A ΔV of 8,000 fps and a specific impulse (Isp) of 300 sec will suffice for such a mission.

The grand tour missions present only a "single opportunity"—spread over two or three years—for those of us now alive—and since it is still some three or four years off, years the Soviet Union is likely to program many other missions in the interim period. Of course, there will be continued missions to explore Venus. The search-for-life experiment on Mars is of great interest to the Soviet scientists and they still would like to be number one in determining the answer to that question. The 1973 opportunity yielded only failures in Soviet attempts for a soft landing on the Martian surface for the major purpose of searching for life.

A mission to Mercury will be quite difficult because of the thermal environment, but I fully expect the Soviet Union to attempt such a mission within the next two or three years. I am, in fact, somewhat puzzled that it has not occurred up to now since even their Zond booster can launch a (hard landing) payload of more than 2,600 lb on a trans-Mercury trajectory ($\Delta V_{b.o.}$ 43,000 fps, Isp = 350 sec).[89]

---

89 Staging the fourth (i.e., injection) stage of the Zond booster yields a 6,900 lb payload, enough for a soft landing.

If there is any reasonable element of truth in all of the press-noted and other tales of failures in unmanned Soviet launchings, then perhaps buried in there somewhere are failures of attempts to reach Mercury. There is no explicit information in the unclassified literature to aid in resolving the question as to whether or not such an attempt was ever made. However, enough information has appeared to indicate that the Soviet Union has suffered an inordinate number of propulsion system failures. It is a good example of how an entrenched hierarchy can permit an often-unsuccessful propulsion group to inflict a continuing penalty on an otherwise outstanding team.

During the course of a paper on data transmission from the planets, Dr. Yu. K.. Khodarev of the Soviet Union made the following remarks:[90]

- Return of black-and-white or color photos with 2 by 107 bits per picture is a requirement (for the Soviet Mars-Venus program).
- He prefers relay satellites in 2-7 hour orbits so that a satellite with mutual Earth-Mars visibility occurs 950 seconds per orbit for data retrieval. He also spoke of preferred circular orbits at an altitude of 280 kilometers with inclinations "not near the Martian equator."
- Would like to get 107-108 bits per day from landers (transmission occurs during part of each day).
- Orbital-relay spacecraft transmitters should have an output between 50 and 100 watts for the Mars and Venus explorations.
- A Venus orbiter at a low inclination, high altitude, circular orbit would be visible to Earth for about two months (interference at other times not explained); however, looking

---

90 Congress of the International Astronautical Federation (IAF) Mar del Plata, Argentina, October 5-10, 1969.

at Venus when it is opposite Earth and close to Earth-Sun line is one such situation.

- An orbit at hc = 1500 km would give a period of 2 hours with 600 seconds per orbit available for data transmission. He spoke of this satellite as being the correct one for working with a Venus lander.

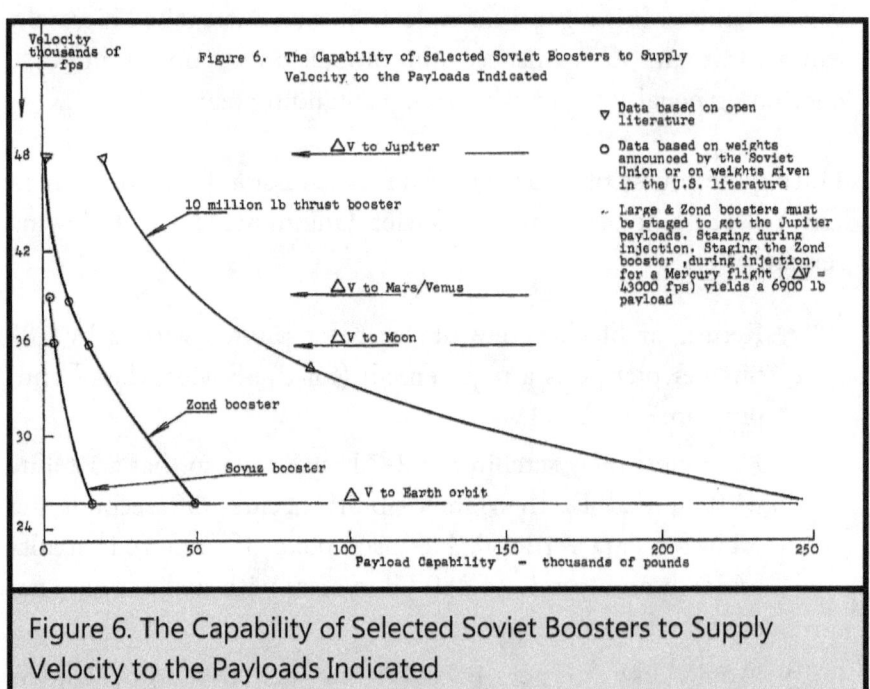

Figure 6. The Capability of Selected Soviet Boosters to Supply Velocity to the Payloads Indicated

Two of these remarks bear further comment. The expressed desire for a Martian orbiter at an altitude of only 280 km indicates that the mission for such a satellite appears to be rather a short-lived one. The lesser gravity of Mars probably yields a greater atmospheric density than is found at the same altitude at Earth, which then leads to a shorter Mars satellite lifetime at 280 km. A surface density of 5 millibars leads to an orbital lifetime of a few weeks for a satellite.

This choice of 280 km appears to be the result of a trade-off between altitude and orbit lifetime for the purpose of conducting photography of the Martian surface. Photography would aid in choosing landing sites for a soft lander. It is entirely possible that the low altitude orbiting satellite is not only a precursor to landing but also may, indeed, include the lander as well. This satellite could be expected to do one of the following after a landing site choice was made:

- Separate and discard orbital photographic equipment and then soft-land the entire package for Mars surface experiments.
- Separate a lander and then raise the remaining orbital payload to a higher orbit for use as a communication relay.
- Reprotect the camera and soft-land the entire package for Mars surface experiments.

The second alternative bears a resemblance to the US 1975 Viking mission. Low-altitude photography alone, seems not quite an entirely likely mission in view of the prior US *Mariner 6* and *7* photographic missions in July and August 1969. The Soviet Union prefers not to repeat missions accomplished by the US; they will always take their own approach to obtain desired information.

That the Soviet Union is contemplating a Venus lander mission bespeaks a substantial step-input in both materials and spacecraft design or, some mighty wishful thinking.[91] The lander idea coupled with thoughts of an orbital relay is interpreted by me as an intent to obtain photos of the Venusian surface via video. The resultant photographs would be some of the most exotic pictures of all time. A camera and associated electronics encased in, say, a

---

91 *Venera 9* and *10*, now part of history, have successfully accomplished such a landing (and its orbital communications relay satellite) in October 1975, together with all the experiments discussed herein.

transparent thermal shield (quartz?) and also protected against the surface pressure of Venus would be an admirable engineering accomplishment.

This kind of experiment would be accompanied, I believe, by a gas chromatograph or mass spectrometer to determine molecular or atomic species. It is also possible that a radiation device like the lunar *Surveyor* spacecraft alpha emitter might be used to examine the Venusian surface. A penetrometer to test the bearing strength of the surface of Venus might also be employed. This could be supplemented by strain gauges on the "legs" of any landing craft. IR devices to determine the surface thermal spectrum would be carried on any Venus spacecraft soft lander.

If the Soviet program really includes soft-landing spacecraft despite the severe surface environment of the surface of Venus, then one might expect assuming a successful lander that three or four launch opportunities later, they will try to bring a specimen of the Venus surface back to Earth. Certainly, such a mission presently sounds exotic and will be difficult to accomplish (there's an understatement!) but a landing success will inevitably lead to a further step in exploration. A sample return is a logical step following soft-landing missions.

Much less difficult than soft landings on Venus is the prospect of floating a payload in the Venusian atmosphere.[92] In so dense an atmosphere it should be relatively easy to suspend several hundred pounds from a balloon. An RTG and batteries should suffice to supply power for such a payload. Solar cells are obviously useless because of the cloud cover. A parabolic antenna could be built into the top of the balloon or the balloon might be radio transparent so that the antenna can be located together with the experiment instrumentation suspended from the balloon.

---

92 I suggested this in the US as early as 1962 but no one was attentive enough to realize the possibilities in such a mission.

It is conceivable that video photos could be obtained in addition to conducting a long interval of atmospheric sampling. If the balloon were at a high enough altitude such that temperatures were not excessive then a search for life investigation would be a worthwhile experimental endeavor. Such a balloon-borne experiment package is not based only on my speculation alone but has origins as well in some discussions published in the Soviet press a few years ago. It would be interesting to learn why such balloon-borne experiments were never carried out since they appear to be much less difficult than the Venus probe experiments that were completed by the Soviet Union and they would be expected to collect data for a considerably longer time—days at least and probably for several weeks.

Although Venus orbiters are another likely mission for the Soviet Union, the kinds of experiments that might be conducted from them are limited because of the very dense cloud cover. Investigation of the makeup of the clouds themselves by examining radio frequency absorption and emission spectra of the clouds would be of interest. The Soviets might try IR tracking and subsequent analysis of burning flares, which have been fired into the clouds from a satellite to aid in determining the cloud constituents, However, it is clear from Khodarev's comments at the IAF conference that the major use of a Venus orbiter would be to serve as a communications relay for a Venus soft lander.

A combined orbiter and lander seem to be a good candidate for a near future mission to Venus. A search from orbit for one of the "cooler" areas of Venus followed by a lander-payload separation and a landing attempt should form the basis for a mission of high interest.[93] The Soviets have expressed on several occasions their desire to investigate the cooler areas of Venus. It may be that instrumentation

---

93 There have been several reports indicating substantial plateau areas of the Venusian surface that are more than 10 kilometers higher than "mean sea level."

may be able to survive for many hours or days on these cooler surface areas and that would provide a measurable improvement in the investigations of Venus. The 1975 launch window opportunity should see the Soviets once again launch spacecraft to Venus. They will probably have the following gross characteristics:

- Spacecraft weight should be 4,000-9,000 lb.
- At least one orbiting communications relay spacecraft will be included.
- A more sophisticated lander will be included. The lander may turn out to be a balloon borne floating-in-the-atmosphere experiment package.
- The Zond booster will be used to launch the spacecraft.
- There will be at least two but more likely three launchings.
- Each spacecraft may consist of an orbiter and lander. Equally likely are separate orbiters and "landers" or "floaters," each launched separately from Earth.[94]

An attempt to conduct a Mars mission failed on March 27, 1969, before the payload reached Earth orbit; this was reported in the *New York Times* of March 28, 1969. The loss resulted from the failure of an upper stage of the booster. Further, an attempt was made, probably successfully, to recover the payload by using the injection stage. While a second attempt may have been made, as in all prior Venus and Mars attempts, there was no information to indicate that it had occurred from any US-based source.

The launch window in May 1971 yielded a new Soviet attempt to reach the surface of Mars. Both of these successful launches—*Mars 2* and *3*—did not complete their primary mission to deliver operational soft-landed spacecraft to the surface of Mars. The landers were, in

---

94 *Venera 9* and *10*, now part of history, have successfully accomplished such a landing (and its orbital communications relay satellite) in October 1975, together with all the experiments discussed herein.

fact, delivered to the surface but the *Mars 2* craft failed completely. The *Mars 3* spacecraft operated for a matter of seconds according to a TASS release. Both *Mars 2* and *3* buses are in orbit around Mars; *Mars 2* has an apoapsis of 15,535 sm and a periapsis of 857 sm, at an inclination of 49°, with a period of 18 hours while *Mars 3* is in an orbit of 118,000 by 932 sm, at an unknown inclination, and with a period of 11 days.

A *Mars 2* component failure after some weeks prevented the receipt of useful data thereafter. *Mars 3* operated for months returning photographs, temperature profiles, IR data, ionospheric, and other data. Each spacecraft weighed 10,231 lb, an impressive number. The landers were probably 2,000 lb each while still attached to the bus and perhaps 750-1000 lb on the surface. They were battery powered and intended to last up to two weeks, no more. Little data describing these spacecraft has been released.

Both of the orbiting buses were intended to act as relay stations for the soft landers. The orbiter has storage capability as well as the tactic of relaying data with only momentary delay. The antenna and its associated mechanism for receiving data from the surface was, presumably, not the same antenna used in the relay-to-Earth procedure. It is reasonable to suppose that it could be used that way in a backup mode. If one of the parabolic antennas on the bus could not be aimed at the lander then even an operational lander on the surface would be useless; sketches showed only whip-type antennas aboard; they are utterly insufficient for broadcast directly to Earth without going through the orbiting relays.

Many prior Soviet interplanetary experiments have been conducted over relatively short periods once the spacecraft have reached their planetary destinations such as *Luna 9* and *13*, and *Venera 4*, *5*, and *6*. However, the possibilities inherent in the discovery of any kind of extraterrestrial life will direct the Mars efforts to much longer times for conducting Mars surface experiments. Despite the parochialism of many researchers on Earth, Martian life may be very different from that on Earth and, indeed, may be somewhat difficult to recognize using the kinds of experiments described in the (US) literature.

The Soviets will probably conduct experiments like [Microbiologist Wolf] Vishniac's "Wolf Trap,"[95] for instance, but one should expect other experimental approaches in view of the oxygen-depleted Martian atmosphere and the consequent effect on any replicating life forms. The Soviet literature, ranging from articles in the daily press (*Pravda, Izvestia*) to science fictioneering (in the magazine *Soviet Life*) and to occasional erudite discussions in technical magazines have dealt with the subject of life on other planets. The various authors have proposed a wide spectrum of life forms, from those involving familiar Earth-type biochemistry to very exotic crystalline life forms (chemistry not explained).

The influence of this variety of thought will manifest itself in the number and kind of experiments that will be conducted on the surface of Mars. These ought to include:

- Stereoscopic photography with resolution to a fraction of a millimeter for a camera-to-object distance of 1-2 meters at that resolution.
- Panoramic photography in the area of the landing location and again out to the local horizon.
- Various biochemistry experiments to ascertain the presence of life within the limits of experiment derived recognition.
- Surface penetrometer and digging mechanisms to determine bearing strength, cohesiveness, granularity, etc.

---

95 Wolf Vladmir Vishniac developed a miniature laboratory, a life-detection instrument.

- Analysis to determine constituents of the Mars surface at the landing location.
- Atmospheric sampler mass spectrometer to determine contents of atmosphere.
- Meteorology experiments to ascertain winds, pressure, density and humidity.

Because of the fact that Mars has been judged to have the greatest chance of extraterrestrial life in the solar system, I expect Soviet efforts to land on Mars and to perform a wide variety of experiments to be an intense and persevering effort. The launch opportunity for 1973 revealed the intensity of their efforts, particularly since it presented the last certain opportunity for them to be number one in this search for life endeavor.[96]

Since the Soviet space program is often full of surprises, I look forward to a Mercury probe in the near future. There have been some profound questions on the nature of a presumed atmosphere on Mercury since the side facing away from the Sun has been measured at a higher temperature than theory indicates it should have. One of the explanations offered assumes the presence of some heavy gases, such as argon and krypton, and consequent convection from the hot side of Mercury. I would add here that there may be continued replenishment of such heavy gases from the Sun. These heavy gases are known to be ejected from the solar surface; they certainly could be captured by the solar-facing side of slowly rotating Mercury and thus account for a convective atmosphere subsequently heating the rearward face of the planet.

Because the planet Mercury is close to the Sun, and has been exposed to relatively intense streams of gases and plasma with

---

96 Four spacecraft were launched for this opportunity but all failed: none reached the surface in operating condition; three missed the planet. A real disaster for the program.

entrapped magnetic fields, there may well be some very different mineral formations and surface material to be found on Mercury.

The scientists and engineers in the Soviet space program are as aware of these phenomena as I am; a perusal of their literature reveals as much. They are also aware of the difficulties involved in placing a spacecraft in orbit around Mercury, in protecting it in that orbit from thermal damage, and in attempting to collect information about Mercury while in that orbit. It appears to me then, that:

- A Mercury probe will be a soft lander.
- The probe will be thermally protected while en route to Mercury by a reflecting shield on the Sun side of the spacecraft backed up by insulation materials.
- The reflective shield will be jettisoned when the probe reaches that area protected by Mercury's umbra.
- After soft landing the payload will be able to broadcast telemetry so long as it remains on the dark side of Mercury.

A trajectory and landing offering the above characteristics should permit protected broadcasting (from the Sun's rays) for several weeks.

The payload should contain some combination of a surface penetrometer, mass spectrometer, surface analyzer (like the gamma emitter on *Luna 13*), gamma ray detector, alpha detector, proton detector, etc., and a seismometer. A video camera will constitute a basic and first-priority piece of equipment.

Since the Zond booster in its present configuration is not capable of injecting a Mercury-bound soft-landing spacecraft, there are two alternatives that may be considered by the Soviets:

- Rendezvous in Earth orbit two components; a Zond booster-launched propulsion module and a payload launched separately by a Soyuz booster, thus obtaining an injected Mercury spacecraft of 5000-7000 lb. The rendezvous procedure would follow the *Cosmos 212/213* technique, or
- Use their 10 million lb thrust booster to send off either a pair of up to 10,000 lb spacecraft or a single 20,000 lb spacecraft.

Both of these would be the basis for a very impressive mission to Mercury.

Soviet launchings to the planets Jupiter or Saturn can occur in one of three ways using available Soviet launch vehicles:

- Use of the Zond booster whose capability for a Jupiter launch is currently limited to 1,500 lb.
- Use of the Zond booster to launch a propulsion module into Earth orbit and subsequent rendezvous with a separate payload launched by a Soyuz booster into Earth orbit in the manner of the *Cosmos 212/213* unmanned automatic rendezvous. A 5,000 to 6,000 lb Jupiter payload is possible with this combination.
- Use of their new 10 million lb thrust, Saturn V class booster. At this writing this booster has not had a successful flight.

For the Soviet Union, 1,500 lb to Jupiter hardly seems worthwhile and I do not expect launchings to Jupiter or beyond to be attempted with the Zond booster alone. Rendezvous has not been demonstrated except by Soyuz spacecraft or its prototypes; *Cosmos 186/188* and *Cosmos 212/213*. That makes the second possibility somewhat remote but not dismissible. The Soviet Union makes every endeavor and perseveres in using its capabilities to the fullest extent. Since their large booster failed in its first known flight attempt they may try the rendezvous technique to get a Jupiter payload en route in the mid-1970s.

The basic problem they face in getting close to Jupiter in orbit is the large velocity increment that must be removed from any Jupiter probe in order to orbit reasonably close to the planet. Of course, for a first photographic look at Jupiter, an orbit many radii out from the planet's surface should suffice. Photography of Jupiter at a distance of about 2,500,000 miles should produce some spectacular pictures. This is at a point where the planet subtends 2° of arc, about the same arc subtended by the Earth at the distance of the Moon. Such photography can certainly be expected as initial information from any Soviet Jupiter mission.

The very powerful microwave emissions from Jupiter's radiation belt will also be an early source of investigation. Both particles and waves would be subjects for investigation. Radiometry examination of Jupiter in the IR and ultraviolet spectra will be used by the Soviets and scanning should commence at a distance of about 3-4 million miles. Since Ganymede is some 665,000 miles from the surface of its primary, photography of it (Ganymede) can be expected on those missions whose trajectories are calculated to fly by that moon. The same comments can be made with respect to other Jovian moons, particularly Callisto.

Now as the Soviets have undoubtedly recognized the orbit of Ganymede is close enough to Jupiter to make a fine vantage point for both radiometric and photographic examination of that planet. However, a circular orbit at the distance of Ganymede is a very demanding mission.

An expected arrival velocity near Jupiter is 52,400 fps. But Ganymede's orbit velocity assuming a circular orbit is 34,640 fps so that a reduction of 17,760 fps is necessary for a spacecraft to achieve an orbit around Jupiter at the distance of Ganymede. This is beyond the booster capability of the Soviet Union and will remain so for some time.

Therefore, we can expect the Soviet probe to have an eccentric orbit around Jupiter whose apogee will be far beyond the orbit of Ganymede. The arrival velocity then will have to be such that the initial velocity decrement is only about 8,000 fps. This will permit them to orbit a payload of approximately 8,000 lb if the entire payload including retro propulsion at arrival is 20,000 lb (again Isp 350).

These values are arbitrary but were selected to yield Soviet "thoughts" on what constitutes a substantial useful payload in an orbit around Jupiter that could include, in addition to the experiment payload, sufficient weight for a power source able to supply 1 kilowatt (a reactor or an RTG) for some weeks. I expect a parabolic antenna of at least 10 meters diameter to be used in the spacecraft's communications system because of the huge distance between the Earth and Jupiter. The 2.4 meter antennas used on the *Venera 4, 5, and 6* spacecraft while en route to Venus are completely insufficient for the Jupiter mission.

Though the Soviet space program managers would dearly like to land on one of the satellites of Jupiter, they are very limited by the capability of their stable of boosters. Even at the distance from Jupiter of Callisto the velocity demands are so high that retro into a fairly near orbit of that moon, say at an altitude of 5000 kilometers (about 3100 statute miles), is almost beyond Soviet capability.[97]

---

97 The derivation of this number is too complex for inclusion in this book. Suffice it to say it was derived using information in the *NASA Planetary Handbooks*.

If a Ganymede or Callisto landing is to be accomplished by the Soviet Union it will be necessary for them to be aided by an atmosphere of the selected moon. Retro capability. Unaided by an atmosphere will not supply sufficient velocity for a Soviet spacecraft to make a soft landing on either Ganymede or Callisto.[98]

A flyby mission of Jupiter and one or more of its moons is within the limits of Soviet capability if and when their large booster becomes operational.12 The possibility of a near orbit mission to Jupiter or one of its moons may be possible by the Soviet Union—although this presently seems rather remote under the following special circumstances:

- A flight is made to Jupiter or Callisto or Ganymede using their large booster to inject a 20,000 lb spacecraft on a Jupiter bound trajectory as previously described.
- Retro velocity of approximately 8,000 fps is applied to place the spacecraft in a wide orbit around Jupiter or around a selected Jovian moon.
- Some experiments are performed, photographs taken, and the data transmitted to Earth.
- A set of plasma jets are turned on in such a manner and for a long enough period so as to cause the orbit to spiral inward thus bringing the spacecraft closer to the planet or closer to one of its selected moons.
- When a suitable altitude is attained the plasma jets are turned off.
- Experiment procedures are reinitiated for a desired interval.

---

98 It is important to point out that if high-energy propellants (hydrogen/oxygen) are introduced into the Soviet space program all of these capabilities will then be available to them.

Chapter 13: The Future

This last technique is suggested since the Soviet Union has demonstrated the use of plasma jets, for attitude control, on three early spacecraft, *Zond 2*, *Voskhod 1*, and *Zond 3*. This method has the advantage of being able to substantially reduce velocity over a long period at relatively little weight because of the inherent high specific impulse, and low mass consumption of plasma jet engines. The application of this propulsion technique is very appropriate when such a system has demonstrated sufficient reliability for a long period.

The Soviet Union has demonstrated the use of plasma jets in October 1964 on *Voskhod 1*, in December 1964 on *Zond 2*, and in July 1965 on *Zond 3*. The intervening time to the present has almost certainly been used by the Soviet plasma jet development group to improve their engines. I am inclined to think that if they have been successful in demonstrating long operating lifetimes the plasma engines will, indeed, be used on a Jupiter mission.

Manned Activity—Earth Orbit

Future activities in Earth orbit for the Soviet Union are well defined, in the sense that they are concentrated around one distinct effort. The central focus is the establishment of a very large, permanently emplaced, multi-manned space station. Because of their supercautious approach to manned flight, progress in this arena will be made in rather small increments. There is another condition, which in parallel with considerations for human life is an overriding factor in the development of a large orbital facility. To date (1975) the largest Soviet payload placed in Earth orbit is some 50,000 lb. This places a severe limit on space station size, and is, in fact, quite inadequate for anything like a large permanent facility.

It is necessary first to digress here in order to define a large orbital facility. Such a station ought to be, at the lower limit, about 500 metric tons—say a million pounds. It should have no less than 50 personnel aboard at all times. Such a space station should have altitude changing capability so as to ensure a very long (many years) lifetime. Such a space station will cost several billion dollars by the time it is placed in orbit and equipped for operation—a good enough reason for the required long lifetime.

The types of activity onboard the station and within its facilities will be limited only by the maturity of available technology. This definition arises from my own deep involvement in original space station design and engineering. Certainly, the space station system can be described in greater detail but these fundamental characteristics suffice to give some bounds on "a large permanent, manned orbital facility." For a look at such a station, see Figure 7.

The Soviet space station *Salyut*, placed in orbit in April 1971, represents a precursor—far removed from the eventual intended facility. It is, however, representative, within a few thousand pounds, of their present capabilities. The idea of orbiting many such payloads—it weighs some 42,000 lb—and docking them in a stable configuration is possible, but is not overly fruitful. In such an assembly a substantial portion of the hardware in orbit, as well as the design efforts that preceded it, would necessarily be devoted to the rendezvous and docking process. Add to that the need for hardware for making the docked joint enduring, properly sealed, and allowing for passageways through the attached compartments.

For two or three such assemblies this procedure is not too unreasonable. Beyond that, too much of the effort in design and in placing the hardware in orbit is designated to a use that could be better devoted to onboard instrumentation and other equipment. A really large station would also have large units rendezvoused and joined, but the number of pieces of hardware designated for such purposes would be small when compared to that used in making up a station from compartments as small as the *Salyut*.

For a million-pound station: if the station is composed of *Salyut*-sized compartments, there are 18 docking interfaces, whereas, with 250,000 lb compartments (based on use of a large booster) there are only 6 such interfaces. This brief discussion excludes the added difficulties and hazards that would be associated with the large number of docking procedures, joining, sealing, and allied processes that go hand-in-hand with the *Salyut*-sized compartments.

The crux of the matter is the lack of a very large booster capable of launching units each of at least 250,000 lb with which to assemble and form a space station.

Despite all of the remarks about the repetitious hardware and complexity associated with multiple docking, the large space station depicted in Figure 7 was designed so as to be assembled using almost 100 separate units. This idea was proposed at a time when 20,000 lb was a large payload for Earth orbit and is quite workable despite the problems previously noted. The large station is 832 tons—1,664,000 lb.[99]

The Soviets could and might use such a modular, multiple assembly approach if they remain booster payload limited. In the long run, however, this limit ought to be overcome by the availability of either a very large booster or a large payload carrying a reusable shuttle.

---

99 I hold a patent for this design, the first ever issued in the US for a space station.

The Soviet Space Shuttle

This brings us to the next facet of Earth orbit activity. During the course of discussions covering cooperation in space the Soviet officials involved alluded to the fact that they have a shuttle program, as well as their space station program, under intensive research and development. Past history provides the basis for presuming that their shuttle will carry a large payload, certainly no smaller than 50,000 lb and more likely closer to 100,000 lb.

When it is initially introduced the shuttle will probably be announced as carrying a payload well below that scheduled for final configuration. This is typical in the Soviet space program and in the development of so different a delivery system for orbital payloads. In not too long a time the payload will then be scaled upward; rapidly if the program is entirely successful, slowly if much debugging becomes necessary.

Should high-energy chemical propellants or a nuclear rocket propulsion system be introduced for the orbiter stage of the shuttle, then the payload will grow even more rapidly. The introduction of the shuttle will enhance their space station program substantially. It will also permit a huge improvement in military reconnaissance programs—larger cameras ergo better resolution, faster response times from picture taking to evaluation of results, and more frequent missions in general.

Passenger, scientist, and military personnel trips to orbit will become a weekly, if not daily, routine affair. I expect to see a 50-man space station by 1982 and routine use of a shuttle in the same period.

Much closer to us in time will be satellites used for investigation of Earth's resources. I think that rather than performing these tasks with unmanned satellites only, information gathering will include the use of manned spacecraft including both the Soyuz and the shuttle. These satellites will be used in at least the following areas—

oceanography, geography, water studies, geology, assessment of plant life, and atmospheric and other environmentally related facets of life on Earth. Correlation of the activities examined will form a substantial series of investigations.

Earth Orbit—Unmanned

Two of the obvious areas for Soviet activity are in meteorological and communications satellites. Both of these have proved to be of great value to the economy of the Soviet Union—in addition to the military value implied in their use. Placing a satellite in synchronous orbit required the Soviet booster to make a plane change of 46°. That implies a very large velocity addition, separate from the attainment of orbit, of as much as 14,000 fps (that is a worst case assuming it is all accomplished at synchronous orbit altitude). Hence the Zond booster was used for these launchings.

Twice in 1974 such satellites were launched; they constituted the *Cosmos 637* and the *Molnyia 1S*. Both likely test vehicles for the operational satellites to follow.[100] A military follow-on, *Cosmos 775*, was emplaced over the mid-Atlantic as an early warning satellite. Use of the Zond booster for this purpose probably means that the satellites are larger than the Intelsat series of satellites and have a greater radiated, useful broadcast power.

Meteorological satellites now look at almost every aspect of weather phenomena. Those of the future will be separated into two classes— those examining global aspects of the weather and those devoted to examination (with intent toward future control) of smaller "local" parts of the weather. The investigations will get detailed enough so that a certain amount of weather control will come about based on the information derived from the satellites. There will be

---

[100] The first Statsionar, operational comsat was launched on December 22, 1975, from Tyuratam.

synchronous weather satellites so that the Soviet Union will have constant information on the macroscopic (large) flow of weather patterns across their country.

Scientific satellites will probably get fewer but much larger and will be aimed at very specific problems. There will be optical and radio astronomy satellites that will "see" many times farther into the universe than comparable land-based instruments.

Cosmic ray telescopes and related devices will be launched to focus on the puzzling and immense energy sources in the distant part of our and other galaxies. If ever their secrets are understood the implications for energy production on Earth will be such as to forever banish the present concern over future sources and techniques for energy production to satisfy our civilization's unending appetite for this commodity. Topside sounders (for ionospheric studies) will collect enough data so that we will make use of the properties of the ionosphere to enhance radio communication rather than contend with the nuisance qualities it affords us now.

Looking farther downstream, I can see the placement of communications satellites in planetary orbits for several purposes. The most obvious will be as interplanetary beacons for Soviet cosmonauts visiting other planets serving as their communications relays and as navigation aids. They may even, in time, be waystations with emergency supplies for interplanetary crews. Such satellites of the solar system will serve another purpose; they will warn Earth of the approach of large pieces of interplanetary debris that may be on an Earth-collision trajectory.

To the Planets—Manned

For as far downstream as I care to look presently, the main future thrust, away from Earth, will be aimed at visits by Soviet cosmonauts to two planets, Venus and Mars. This will not occur until unmanned investigations bring back some basic data on those planets. That process has been long underway. The trip to Venus will be good practice for the Mars trip. There will not be a manned Venus landing in the foreseeable future but a trip to Venus orbit is reasonable.

Such a spaceship will be laden with unmanned probes of many kinds for study of the Venus atmosphere and surface and possibly a surface sample retrieval or two. The Mars expedition will be far more complex. Enough has been written elsewhere, or speculated about, that I find no need to detail what studies the Soviets will conduct on that tempting planet. It is necessary to state, conclusively, that the pap fed to the public by some writers on the crew and spacecraft sizes ought to be "shot down."

No crew is going to the planets in any of the spacecraft that now exist or in anything like them. Nor is the Soviet Union going to send just three or four men on so challenging an expedition. With the best-known technology, the smallest spacecraft—all-up weight—that I can see making the trip to Mars will be 1,000,000 pounds.[101] The smallest crew that I can envision as being useful will be 10 men and women. I would surmise that the all-up spacecraft could be as much as five times as large and the crew possibly three times as

---

101 Clearly, this excludes all major propulsion units.

Figure 7. Space Station system patented by the author

large. Sending a crew to a planet some 40,000,000 miles away on a round trip of some two years duration is not a trip to the South Pole. There will be few opportunities, if any, for the crew to change their collective minds, abandon the expedition, and come straight home quickly!

After more than two weeks into the Mars trajectory, the difficulties in aborting the mission are either insurmountable or place a tremendous burden on the payload placed in Earth orbit initially. A Mars trip is a great risk and great nations and their people are always willing to make such efforts when they have the ability to do so. Commensurate with the era beginning in 1986, the Soviet Union should be ready to undertake the greatest expedition in the history of man. It is entirely possible that the Mars trip could be preceded by a less than one year, roundtrip to Venus orbit and back.

It would seem that the opportunities for Soviet manned exploration of other bodies in the solar system ought to be strongly focused on our own puzzling natural satellite, the Moon. It still enshrouds mysteries enough to keep teams of scientists and engineers engaged in lunar research for years to come if only the nearside is explored and studied. Add to that, the remote farside and its endless highlands, sufficient reason is provided for the establishment of a permanently manned lunar base. I don't doubt that such a base is in the minds of Soviet space research scientists. Twenty-three Luna series spacecraft plus several more attempts that were never identified as lunar probes, have provided more than enough evidence of an overwhelming interest in the Moon.

Whenever I am reminded of the motion picture, *2001: A Space Odyssey*, it occurs to me that based on the present state of progress and interest displayed in space exploration the roles of the US and the Soviet Union depicted in the film are, in a realistic extrapolation of the present situation, quite likely to be reversed. Of course, one would hope that in any real-life occurrence of such an eventuality the foolish Cold War aspects of life that found their way into the film would be nonexistent. Events of mid-year 1975 show that cooperation will, indeed, be the theme of the future.

Because a good deal of lunar exploration has occurred—though some might call it primitive so far—and because Mars still stands out as the prime arena in the solar system upon which to find extraterrestrial life, it isn't farfetched to expect that a Soviet manned Mars expedition would precede the establishment of a permanent lunar base.

All of this prophesy for prolonged manned exploration of the Moon, like that surmised for planetary expeditions, is keyed to the appearance of a very large booster and probably a reusable space shuttle. Without such vehicles the exploration would be extremely difficult, if not altogether impossible, and would be, to put it mildly, inordinately expensive.

Ergo, if I am to predict these exotic future explorations then I must also predict beforehand the operational use of these spacecraft and boosters. An additional and immeasurably advantageous ancillary spacecraft would be a so-called space tug, used to perform various services on orbit. The latter include docking operations, refueling, servicing of many kinds, and personnel and cargo transfer functions.

Before another decade has passed in the still embryonic Space Age, all these things will come to pass. The greatest adventures in the history of man shall then commence and they will open the door to infinity a little bit wider.

Appendix A: A Calculation for Eccentricity

Appendix A

The equation for the determination of the eccentricity of a satellite orbit.

There is no intent here to rigorously derive such an equation but rather to assume the correctness of two basic equations from orbital mechanics and then arrive at a result simple enough for anyone to use.

Thus, from the mathematics of orbital mechanics we learn that r_p and r_a the radial distances from the center of the Earth to the low point (perigee) and the high point (apogee) of an orbit are related to the eccentricity in the following manner,

$$r_p = a(1 - e) \quad \text{and} \quad r_a = a(1 + e) \tag{1}$$

where, a is the semi-major axis of the orbit, and

e is the eccentricity

Since the radial distances are measured from the center of the Earth they may be further expressed as,

$$r_p = R_o + h_p \quad \text{and} \quad r_a = R_o + h_a \tag{2}$$

where, R_o is the mean radius of the Earth, and

h_p, h_a are the distances from the Earth's surface to perigee and apogee of the orbit respectively.

If we add the equations in (1) we get,

$$r_p + r_a = 2a \tag{3}$$

If we subtract the equations in (1) we get,

$$r_a - r_p = a(1 + e) - a(1 - e) = 2ae \tag{4}$$

Substituting equations (2) and (3) into (4) yields

$$(R_o + h_a) - (R_o + h_p) = 2ae = (R_o + h_a + R_o + h_p) e \tag{4'}$$

Solving then for e brings us to the desired result,

$$e = \frac{|h_a - h_p|}{2R_o + h_a + h_p} \tag{5}$$

Hence, just knowing the perigee and apogee of an orbit, e can be calculated easily.

Appendix B:
Letters on
Vostok 3 and 4

Missiles & Rockets,
3 September 1962

Vostoks III & IV

To the Editor:

I would like to add some useful information to the continuing debate that has arisen as a result of the Russian demonstration of guidance, control and multiple launching capability in the prolonged flights of Cosmonauts Nikolayev and Popovich. Because of the (relatively) low perigees, the orbit parameters for *Vostok III* and *IV* changed rather rapidly; most calculations made are necessarily somewhat inexact. Nevertheless, for the present commentary such variations are trivial and may be neglected. Consider the following orbital parameter values published in *Izvestia* and *Pravda* of Aug. 12 and 13:

	Vostok III	*Vostok IV*
Orbital Period	88.2 min.	88.3 min.
Orbit Plane Inclination	64° 59'	64° 57'
Apogee	141.42 s. mi.	145.9 s. mi.
Perigee	109.8 s. mi.	110.54 s. mi.

A plane (more or less) normal to the *Vostoks'* velocity vectors, cutting both orbit planes and passing through the Earth's center yields the diagrams in the accompanying figure. This indicates the geometric relationship between the spacecraft during their orbital sojourn. As can be seen in the figure, the volume that contained both spacecraft had a varying (approximately) trapezoidal cross-section defined by points ABCD. At two points, the nodes of these orbital planes, the horizontal dimension (e.g., AB and CD) go to zero, leaving only an altitude variation to consider.

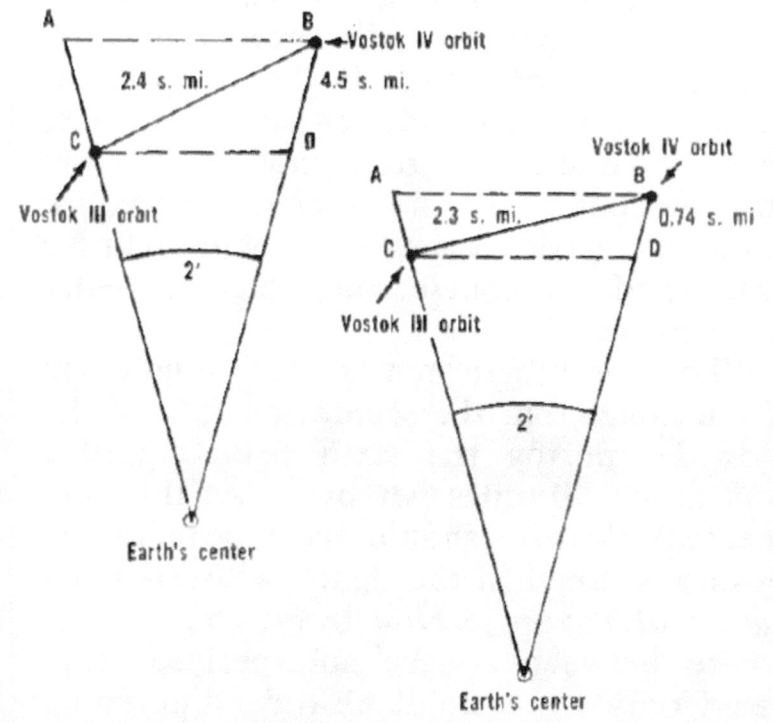

These particular points should be noted:

A. The maximum altitude difference between the two spacecraft occurred at apogee; this was 4.5 statute miles.

B. The minimum altitude separation was at perigee; this was 0.74 statute miles.

C. The angular separation of the orbits was 2 minutes of arc, resulting in a maximum distance between the spacecraft of 2.4 statute miles at apogee and a minimum distance of 2.3 statute miles at perigee.

Consequently, it can be seen that the slot that the *Vostoks* traveled in was never more than 2.4 miles wide and never more than 4.5 miles in height. This indicates a high capability for accomplishing the desirable procedure of rendezvous. Moreover, since the separation distance along the orbital path, noted early in the group flight, during the sixth orbit, was something less than 75 miles, it is reasonable to assume that a rendezvous and docking procedure could have been carried out with relative ease (and, of course, may have actually taken place).

The parenthetic remark is strengthened if one notes that the separation rate of the *Vostoks*, during the sixth orbit sighting, was about 50 miles per orbit. At this rate the two *Vostoks* should have been in the positions noted in the figure with *no separation along the orbital track* (e.g., somewhere between apogee and perigee separated only by a small altitude differential and a small horizontal distance) at about the interval midway between the fourth and fifth orbits. Thus the spacecraft either changed positions during this interval or they were quite close (docked?) and then commenced separation, using an impulse to place *Vostok III* in the lead for the remainder of the flight.

SAUNDERS B. KRAMER
Senior Member, American Astronautical Society
Fellow, British Interplanetary Society
Sunnyvale, Calif.

Appendix C: The Injection Velocity for *Verena 5*

Appendix C

The calculation for the velocity required for the injection into an interplanetary trajectory of the Venus probe, Venera 5.

The spacecraft was launched on January 5, 1969 and the NASA Planetary Maps show a judicious selection by the Soviet Union for the velocity required for injection on that day was minimal.

At an average altitude of some 115 sm the escape velocity is 36,175 fps while the Earth satellite velocity for that altitude is 25,580 fps.

Now the total velocity at injection, ΔV_T, is found by using the equation,

$$\Delta V_T = \sqrt{V_e^2 + V_{hp}^2} \qquad (1)$$

where, V_e is the escape velocity and V_{hp} is called the hyperbolic excess velocity. The latter is the ΔV over and above escape to 'steer' the spacecraft on to the correct course to the selected planetary target. Its calculation is relatively complex and won't be derived here but values are easily obtained from the planetary velocity contour maps made by Dr. Stanley Ross for NASA. These maps yield such velocities as a fraction of the Earth's Mean Orbital Speed (EMOS) around the sun. This value is about 96,100 fps.

For the present case the EMOS fraction was 0.100 or 9622 fps, thus,

$$\Delta V_T = \sqrt{(36175)^2 + (9622)^2} = 37,433 \text{ fps} \qquad (1a)$$

Since the spacecraft, Venera 5, was already in Earth orbit at 25,580 fps, the necessary injection velocity, ΔV_i, was,

$$37,433 - 25,580 = 11853 \text{ fps} \qquad (2)$$

C-2

Using ΔV_i in the standard rocket equation yields the propellant weight (mass) for this maneuver,

$$W_p = (1.2W_p + 2491)(1 - e^{-\frac{11853}{(320)(32.2)}}) \qquad (3)$$

where, the specific impulse, I_{sp} = 320 lb-sec/lb
the acceleration of gravity is 32.2 ft/sec^2
2491 lb is the payload (spacecraft) weight
1.2W_p accounts for the propellant weight and 0.2 is the fraction that accounts for tankage to contain the propellant plus rocket engine, controls, etc.

The solution to equation (3) is W_p = 8940 lb
Since the Soviet Union announced the time of burning for the injection rocket engine as Δt = 228 sec, the burning rate can be obtained as,

$$(8940 \text{ lb})/(228 \text{ sec}) = \dot{w} = 40 \text{ lb/sec} \qquad (4)$$

Now thrust may be defined from the relationship

$$I_{sp} \dot{w} = T$$

so that, $\qquad$ (320 lb-sec/lb)(40 lb/sec) = 12,800 lb $\qquad$ (5)

is the thrust of the injection rocket engine.

Appendix D: The Burn-Time for the *Cosmos 359* Aborted Flight (nee-*Verena 8*)

Appendix D

Determination of the burning time of the planetary injection engine for Cosmos 359 (nee Venera 8), a Venus probe that failed to leave Earth orbit.

As with prior Soviet Venus probes the launching of Cosmos 359 took place when the required injection velocity was, as usual, at a minimal value. We shall need the injection velocity, ΔV_i for this exercise. Using the information in Appendix C we find that EMOS was 0.103, barely larger than used for Venera 5 in the launch window nineteen months earlier. Therefore,

$$\Delta V_T = \sqrt{(35975)^2 + (9910)^2} = 37,315 \text{ fps} \quad (1)$$

Planetary injection normally would have occurred at the altitude of booster burnout (where the third stage finally orbited) so ΔV_i is determined therein. Also, the Earth satellite ΔV at 160 sm where the stage was is 25,436 fps so that,

$$\Delta V_i = 37,315 - 25,436 = 11,879 \text{ fps} \quad (2)$$

It is now required to determine the ΔV necessary to take the payload to its final orbit apogee from the lower orbit of the booster third stage; that is, to go to apogee at 565 sm from the circular orbit at 160 sm. Sufficient ΔV added at the lower altitude establishes a perigee there with apogee at the (for our case, known) upper altitude. We know the end points, it remains to use the proper equation to determine the ΔV. From orbital mechanics we have,

$$\Delta V = V_p - V_c = [(2g_o R_o^2/(r_2 + r_1))]^{1/2} [r_2/r_1]^{1/2} - [g_o R_o^2/r_1]^{1/2} \quad (3)$$

D-2

where, g_o is the acceleration due to gravity
r_2 is the apogee measured from the center of the Earth
r_1 is the newly established perigee (at 160 sm), also measured from the center of the Earth and equals 4119 sm
V_p is the newly established perigee velocity
V_c is the circular orbit satellite velocity at 160 sm

Substitution of the values in equation (3) yields,

$$\Delta V = V_p - V_c = 26019 - 25436 = 583 \text{ fps}$$

Now changes in orbit inclination are costly in terms of propulsion; one degree change 'costing' some 400 fps. Since the change was 0.6 degrees the necessary ΔV was 240 fps.

Taking the root mean square of these two ΔV values yields,

$$\sqrt{(583)^2 + (240)^2} = 630 \text{ fps}$$

When the Venera injection stage (actually a fourth stage on the booster/payload combination) burns for its full interval it adds a velocity of 11,879 fps in 244 seconds. This burn time was announced for the injection of Venera 7 by the Soviet Union. Because of the laws of celestial mechanics it can't vary but infinitesimally for a similar launch five days later ; consequently, now knowing the actual velocity added by the injection stage, the total burn time for the stage, and the total velocity intended for the event, a ratio will reveal the burn time for the aborted event; thus,

$$\Delta t_{abort} = (630/11879) \times (244 \text{sec}) = 12.94 \text{ sec}$$

Appendix E: NASA Prediction Bulletin

NASA PREDICTION BULLETIN

NASA 51004

NASA GODDARD SPACE FLIGHT CENTER, CODE 512, GREENBELT, MD. 20771
ISSUE DATE: December 3, 1981

```
BLTN   72 ELEM   72 OBJ 12871    T: 616              ; IN 3 PARTS. PART I
1 12871J            81336.21079138  .00052137   00202-0   00020-2 0 00721
2 12871   82.9516 289.8945 0038149 113.2495 247.2716 15.29263180  9811
                    FOR OFFICIAL USE ONLY. THIS PREDICTION SHOULD
                    NOT BE USED FOR ANY PRECISE SCIENTIFIC ANALYSIS.
PART II S-N EQUATOR CROSSINGS.
   REV  TIME Z  LONG W      REV  TIME Z  LONG W      REV  TIME Z  LONG W
30 NOV 81
   962 2313.23 126.97
 1 DEC 81
   963   47.46 150.65       964  221.69 174.34       965  355.92 198.22
   966  530.15 221.70       967  704.37 245.39       968  838.60 269.27
   969 1012.83 292.75       970 1147.26 316.44       971 1321.28 340.12
   972 1455.51   3.80       973 1629.74  27.49       974 1803.96  51.17
   975 1938.19  74.85       976 2112.41  98.54       977 2246.64 122.22
 2 DEC 81
   978   22.87 145.90       979  155.09 169.58       980  329.31 193.27
   981  503.54 216.95       982  637.76 240.63       983  811.99 264.32
   984  946.21 288.20       985 1120.43 311.68       986 1254.66 335.36
   987 1428.88 359.04       988 1603.10  22.73       989 1737.33  46.41
   990 1911.55  70.09       991 2045.77  93.77       992 2219.99 117.46
   993 2354.21 141.14
 3 DEC 81
   994  128.43 164.82       995  302.65 188.50       996  436.87 212.18
   997  611.10 235.86       998  745.32 259.55       999  919.53 283.23
  1000 1053.75 306.91      1001 1227.97 330.59      1002 1402.19 354.27
  1003 1536.41  17.95      1004 1710.63  41.63      1005 1844.85  65.32
  1006 2019.07  89.20      1007 2153.28 112.68      1008 2327.50 136.36
 4 DEC 81
  1009  101.72 160.04      1010  235.94 183.72      1011  410.15 207.40
  1012  544.37 231.08      1013  718.58 254.76      1014  852.80 278.44
  1015 1227.22 302.12      1016 1201.23 325.80      1017 1335.45 349.48
  1018 1509.66  13.16      1019 1643.88  36.84      1020 1818.09  60.52
  1021 1952.30  84.20      1022 2126.52 107.88      1023 2300.73 131.56
 5 DEC 81
  1024   34.94 155.24      1025  209.16 178.92      1026  343.37 202.60
  1027  517.58 226.28      1028  651.79 249.96      1029  826.01 273.64
  1030 1000.22 297.32      1031 1134.43 321.00      1032 1308.64 344.68
  1033 1442.85   8.36      1034 1617.06  32.04      1035 1751.27  55.72
  1036 1925.48  79.40      1037 2059.69 103.08      1038 2233.90 126.76
 6 DEC 81
  1039    8.11 150.43      1040  142.32 174.11      1041  316.53 197.79
  1042  450.74 221.47      1043  624.95 245.15      1044  759.15 268.83
  1045  933.36 292.51      1046 1107.57 316.18      1047 1241.78 339.86
  1048 1415.98   3.54      1049 1550.19  27.22      1050 1724.40  50.90
  1051 1858.60  74.58      1052 2032.81  98.25      1053 2227.01 121.93
  1054 2341.22 145.61
 7 DEC 81
  1055  115.42 169.29      1056  249.63 192.97      1057  423.83 216.64
  1058  558.24 242.32      1059  732.24 264.00      1060  906.45 287.60
  1061 1240.65 311.35      1062 1214.85 335.23
PART III. REDUCTION TO OTHER LATITUDES AND HEIGHTS FOR REV 1012
  LAT MINUTES      L        HT       LAT MINUTES      L        HT
   N   PLUS      CORR      KILOM      S   PLUS     CORR      KILOM
  SN   0   0.00    2.32    487.7I    NS   0   46.83   191.77   473.5
  SN  10   2.62  359.42    492.6I    NS  10   49.45   191.19   479.7
  SN  20   5.24  358.75    478.4I    NS  20   52.07   190.52   487.2
  SN  30   7.87  357.91    475.2I    NS  30   54.70   199.68   495.0
  SN  40  10.50  356.72    472.9I    NS  40   57.35   188.50   503.3I
  SN  50  13.14  354.88    471.2I    NS  50   60.03   186.65   511.3I
  SN  60  15.81  351.68    469.9I    NS  60   62.73   183.48   518.6I
  SN  70  18.54  344.93    468.8I    NS  70   65.50   176.73   524.6I
  SN  80  21.58  321.25    467.6I    NS  80   68.59   153.07   529.1I
   N  PT  23.45  275.89    466.7     S  PT   72.49   127.72   530.5I
  NS  80  25.31  232.53    465.8     SN  80   72.40    62.37   530.9I
  NS  70  28.35  206.86    464.3     SN  70   75.49    38.71   529.2I
  NS  60  31.07  200.10    463.1     SN  60   78.26    31.96   525.5I
  NS  50  33.73  196.92    462.3     SN  50   80.97    28.77   520.4I
  NS  40  36.37  195.26    462.2     SN  40   83.65    26.94   514.1I
  NS  30  38.99  193.87    463.1     SN  30   86.31    25.76   507.2I
  NS  20  41.61  193.22    465.2     SN  20   88.95    24.92   500.6I
  NS  10  44.22  192.36    469.6     SN  10   91.59    24.26   493.6I
  NS   0  46.83  191.77    473.5     SN   0   94.22    23.68   487.6I
```

Acronyms

AFB	Air Force Base
AFSOC	Air Force Special Operations Command
AFSPC	Air Force Space Command
AFWA	Air Force Weather Agency
AOC	Air Operations Center
AOR	Area of Responsibility
AQI	Al Qaeda in Iraq
ATO	Air Tasking Order
AWACS	Airborne Early Warning and Control System
BFT	Blue Force Tracking
BGAN	Broadband Global Area Network
C2	Command and Control
C4I	Command, Control, Communications, Computers, and Intelligence
CALCM	Conventional Air-Launched Cruise Missiles
CAOC	Combined Air Operations Center/Combined Air and Space Operations Center
CAP	Combat Air Patrols
CCAFS	Cape Canaveral Air Force Station

CFACC	Combined Forces Air Component commander
CINCNORAD	Commander in Chief NORAD
COCOM	Combatant command
COMAFFOR	Commander of Air Force Forces
COMCAM	Combat Camera
COMSPACEAF	Commander Space Air Forces
CONUS	Continental United States
CSAR	Combat Search and Rescue
DCA	Defense Communications Agency
DDMS	DoD Manned Spaceflight Support Office
DIRSPACEFOR	Director of Space Forces
DISA	Defense Information Systems Agency
DMSP	Defense Meteorological Satellite Program
DoD	Department of Defense
DSCS	Defense Satellite Communications System
DSP	Defense Support Program
EHF	Extremely high frequency
EMSS	Enhanced Mobile Satellite Service
ETCS	Expeditionary Tactical Communications System
FLTSATCOM	Fleet Satellite Communications
FBCB2	Force XXI Battle Command Brigade-and-Below
GAN	Global Access Network
GBS	Global Broadcast System
GBS	Global Broadcast Service
GBU-39	Guided Bomb Unit-39

GETS	GPS Enhanced Theater Support
GIS	Geographic Information System
GMLRS	Guided Multiple-Launch Rocket System
GPS	Global Positioning System
ID	Infantry Division
IPT	Integrated product team
IR	Infrared
ISC2	Integrated Air and Space Command and Control
ISR	Intelligence, Surveillance and Reconnaissance
JBS	Joint Broadcast System
JDAM	Joint Direct Attack Munitions
JFACC	Joint Force Air Component Commander
JFC	Joint Force Commander
JNWC	Joint Navigation Warfare Center
JPADS	Joint Precision Air Drop System
JSpOC	Joint Space Operations Center
KSC	Kennedy Space Center
LDR	Low-Data-Rate
LGB	Laser-Guided Bombs
LIDAR	Light Detection and Ranging
MACSAT	Multiple Access Communications Satellites
Mbps	Megabits per second
MCS	Master Control Station
MDR	Medium-Data-Rate
METSAT	Meteorological satellite

MILSATCOM	Military Satellite Communications
MPDS	Mobile Packet Data Service
NASA	National Aeronautics and Space Administration
NATO	North Atlantic Treaty Organization
NGS	National Geodetic Survey
NOAA	National Oceanic and Atmospheric Administration
NORAD	North American Aerospace Defense Command
NSSI	National Security Space Institute
OIF	Operation Iraqi Freedom
OLS	Operational Linescan System
PDA	Personal Digital Assistants
PNT	Positioning, Navigation, and Timing
RF	Radio-frequency
RFID	Radio Frequency Identification
SATCOM	Satellite communications
SBIRS	Space-Based Infrared System
SCA	Space coordinating authority
SDB	Small-Diameter Bomb
SEAS	Sun Earth Acquisition Sequence
SEP	Spherical Error Probable
SHF	Super-high-frequency
SIDC	Space Innovation and Development Center
SIPRNET	Secret Internet Protocol Network
SLEP	Service Life Enhancement Program
SLGR	Small, Lightweight, GPS Receiver

SMART-T	Secure Mobile Anti-Jam Reliable Tactical Terminals
SOC	Space Operations Center
SOPS	Space Operations Squadron
SpaceWOC	Space Weather Operations Center
SPOT	Satellite Pour l'Observation de la Terre
SPS	Standard Positioning System
SSO	Senior space officer
STO	Space Tasking Order
SW	Space Wing
SWC	Space Warfare Center
SWO	Space Weapons Officers
TacMedCS	Tactical Medical Coordination System
TBMCS	Theater Battle Management Core System
TENCAP	Tactical Exploitation of National Capabilities
TSC	Theater Support Cell
UAV	Unmanned aerial vehicle
UFO	UHF Follow-On
UK	United Kingdom
USCENTCOM	United States Central Command
USCINCSPACE	Commander in Chief of USSPACECOM
USSPACECOM	U.S. Space Command
USSTRATCOM	United States Strategic Command
VAB	KSC Vehicle Assembly Building
WGS	Wideband Global SATCOM
WTC	World Trade Center

About the Publisher

The SPACE 3.0 Foundation

SPACE 3.0 is a 501(c)(3) charitable foundation with a mission to preserve space history and empower space entrepreneurs and visionaries.

To learn more about the Foundation, our activities, or to learn how to get involved, please visit: spacecommerce.org

Quest: The History of Spaceflight Quarterly

Published since 1992, *Quest* is the oldest peer-reviewed publication focused exclusively on the history of spaceflight. Each issue features the people, programs, and politics that made the journey into space possible. Written by professional and amateur historians, *Quest* brings you stories and behind-the-scenes insight.

For more information or to order, please visit: spacehistory101.com

www.ingramcontent.com/pod-product-compliance
Lightning Source LLC
Chambersburg PA
CBHW020633230426
43665CB00008B/156

www.ingramcontent.com/pod-product-compliance
Lightning Source LLC
Chambersburg PA
CBHW020651230426
43665CB00008B/385